Preface

Aromatic chemistry, in terms of the production of derivatives of benzene and, to a less extent, other carbocyclic aromatic compounds, is of immense industrial importance and is the mainstay of many chemical companies. Derived products are in general use across such diverse industries as pharmaceuticals, dyestuffs, and polymers.

The aromatic chemistry required by an undergraduate in chemistry, biochemistry, materials science and related disciplines is assembled in this text, which also provides a link to other aspects of organic chemistry and a platform for further study. In line with the series style, a number of worked problems and a selection of questions designed to help the student to understand the principles described are included.

The first chapter discusses the concept of aromaticity, after which there is a description of aromatic substitution reactions. Chapters covering the chemistry of the major functionalized derivatives of benzene follow. A chapter on the use of metals in aromatic chemistry discusses not only the chemistry of Grignard reagents and aryllithium compounds but also the more recent uses of transition metals in the synthesis of aromatic compounds. The penultimate chapter discusses the oxidation and reduction of the benzene ring and the text concludes with the chemistry of some polycyclic compounds.

We have chosen to use the names of chemicals that are in common usage on the basis that students should then be able to read and make use of the chemical literature and also to locate chemicals in the laboratory. Systematic names are given in parentheses at the first appropriate opportunity. Ideally, a student should be able to use both systems interchangeably without difficulty. The RSC website has an Appendix of Common and Systematic Names (http://www.chemsoc.org/pdf/tct/functionalappendix.pdf) to which students are referred. A Further Reading list is also available at (http://www.chemsoc.org/pdf/tct/functionalreading.pdf).

We are grateful to Dr. Mark Heron for his valuable comments on the draft manuscript and to Dr. Alan Jones and Ms. Beryl Newell for their help in preparation of the final manuscript. Mr. Martyn Berry and Professor Alwyn Davies FRS offered advice, encouragement and criticism throughout the preparation of the text which were most appreciated. Mrs. Janet Freshwater of the Royal Society of Chemistry was involved in the project from start to finish and we thank her for her efficiency and guidance. We thank our wives, Annabelle, Margaret and Anita, for their help, patience and understanding during the writing of this book.

J. D. Hepworth, *University of Central Lancashire*
D. R. Waring, *formerly of Kodak Ltd., Kirkby, Liverpool*
M. J. Waring, *AstraZeneca, Alderley Park, Cheshire*

TUTORIAL CHEMISTRY TEXTS

This series of books consists of short, single-topic or modular texts, concentrating on the fundamental areas of chemistry taught in undergraduate science courses. Each book provides a concise account of the basic principles underlying a given subject, embodying an independent-learning philosophy and including worked examples. The one topic, one book approach ensures that the series is adaptable to chemistry courses across a variety of institutions.

TITLES IN THE SERIES

Stereochemistry *D G Morris*
Reactions and Characterization of Solids *S E Dann*
Main Group Chemistry *W Henderson*
d- and f-Block Chemistry *C J Jones*
Structure and Bonding *J Barrett*
Functional Group Chemistry *J R Hanson*
Organotransition Metal Chemistry *A F Hill*
Heterocyclic Chemistry *M Sainsbury*
Atomic Structure and Periodicity *J Barrett*
Thermodynamics and Statistical Mechanics *J M Seddon & J D Gale*
Basic Atomic and Molecular Spectroscopy *J M Hollas*
Organic Synthetic Methods *J R Hanson*
Aromatic Chemistry *J D Hepworth, D R Waring & M J Waring*
Quantum Mechanics for Chemists *D O Hayward*

FORTHCOMING TITLES

Mechanisms in Organic Reactions
Molecular Interactions
Reaction Kinetics
X-ray Crystallography
Lanthanide and Actinide Elements
Maths for Chemists
Bioinorganic Chemistry
Chemistry of Solid Surfaces
Biology for Chemists
Multi-element NMR
Peptides and Proteins
Biophysical Chemistry
Natural Product: The Secondary Metabolites

Further information about this series is available at www.chemsoc.org/tct

Orders and enquiries should be sent to:
Sales and Customer Care, Royal Society of Chemistry, Thomas Graham House, Science Park, Milton Road, Cambridge CB4 0WF, UK

Tel: +44 1223 432360; Fax: +44 1223 426017; Email: sales@rsc.org

TUTORIAL CHEMISTRY TEXTS

13

Aromatic Chemistry

JOHN D. HEPWORTH
University of Central Lancashire

DAVID R. WARING
formerly of Kodak Ltd., Kirkby, Liverpool

MICHAEL J. WARING
Astra Zeneca, Alderley Park, Cheshire

Cover images © Murray Robertson/visual elements 1998–99, taken from the 109 Visual Elements Periodic Table, available at www.chemsoc.org/viselements

ISBN 0-85404-662-3

A catalogue record for this book is available from the British Library

Published by The Royal Society of Chemistry, Thomas Graham House, Science Park, Milton Road, Cambridge CB4 0WF, UK
Registered Charity No. 207890
For further information see our web site at www.rsc.org

Typeset in Great Britain by Wyvern 21, Bristol
Printed and bound by Polestar Wheatons Ltd, Exeter

Contents

1
Aromaticity

Aims

By the end of this chapter, you should understand:

- The structure of benzene
- The concept of aromaticity and its application to various organic systems
- The fundamentals of naming derivatives of benzene

1.1 Introduction

The classification of organic compounds is based on the structure of the molecules. **Aliphatic** compounds have open-chain structures such as hexane (**1**) and can contain single (C–C), double (C=C) and triple (C≡C) bonds. In **alicyclic** molecules, the carbon atoms form a cyclic structure, as in cyclohexane (**2**) and cyclohexene (**3**).

Aromatic compounds are unsaturated cyclic molecules that possess additional stability as a result of the arrangement of π-electrons associated with the unsaturation of the ring system. This book will concentrate on the chemistry of benzene (**4**) and its derivatives and related polynuclear hydrocarbons. Aromatic compounds are also known as **arenes**; they can be **carbocyclic**, indicating that the ring skeleton contains only carbon atoms, or **heterocyclic**, with at least one atom other than carbon in the ring. These heteroatoms are typically N, O or S. Heterocyclic compounds, which can be aromatic or alicyclic, are covered in another book in this series.

Initially, we will look at what distinguishes aromatic compounds from other cyclic molecules and how chemists' understanding of aromaticity has developed up to the present day.

1 2 3 4

The term "aromatic" acknowledges the fact that many fragrant compounds contain benzene rings, whilst the term "aliphatic" is derived from the Greek word for fat or oil. Benzene was isolated in 1825 by Faraday from whale oil and in 1845 by Hofmann from coal tar. Benzene is a colourless, flammable liquid, with a boiling point of 80 °C. It is carcinogenic.

1.2 Structure of Benzene

Based on elemental composition and relative molecular mass determinations, the formula of benzene was found to be C_6H_6. The saturated hydrocarbon hexane has the molecular formula C_6H_{14} and therefore it was concluded that benzene was unsaturated. Kekulé in 1865 proposed the cyclic structure **4** for benzene in which the carbon atoms were joined by alternate single and double bonds. Certain reactions of benzene, such as the catalytic hydrogenation to cyclohexane, which involves the addition of six hydrogen atoms, confirmed that benzene was a ring compound and that it contained three double bonds. However, since benzene did not undergo addition reactions with HCl and HBr, it was concluded that these double bonds were different from those in ethene and other unsaturated aliphatic compounds.

Note that these structures were proposed before the electron had been discovered and before the idea of the covalent bond had been developed.

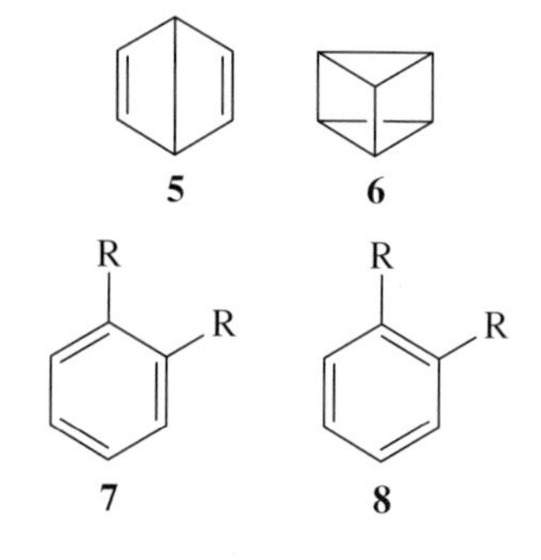

In 1867, Dewar proposed several possible structures for benzene, one of which was **5**. However, in 1874, Ladenburg proved experimentally that all the hydrogen atoms of benzene were equivalent and suggested the prismatic structure **6**.

Kekulé's proposed structure **4** looks more in keeping with our current knowledge of benzene, although it does not explain how the double bonds differ from the aliphatic type. Furthermore, although the two structures **7** and **8** can be drawn for a 1,2-disubstituted benzene, only one such compound exists. Kekulé proposed that the equivalent structures **7** and **8** oscillated between each other, averaging out the single and double bonds so that the compounds were indistinguishable.

1.3 Stability of the Benzene Ring

Kekulé's proposals gained wide acceptance and were supported by the experimental work of Baeyer in the late 19th century, but these ideas did not explain the unusual stability of benzene. This is typified by its chemical reactions, which are almost exclusively substitution rather than the expected addition. Throughout this book there will be many examples of this property. In addition, physical properties such as enthalpies of hydrogenation and combustion are significantly lower than would be expected for the cyclohexatriene structure of Kekulé. The enthalpy of hydrogenation (ΔH) of the double bond in cyclohexene is -120 kJ mol^{-1} and that of cyclohexa-1,3-diene with two double bonds is almost twice that at -232 kJ mol^{-1}. Cyclohexatriene, if it existed, would be expected to have an enthalpy of hydrogenation of three times the value of cyclohexene, a ΔH of approximately -360 kJ mol^{-1}. However, the value for benzene is less exothermic than this comparison suggests, being only -209 kJ mol^{-1}. Thus benzene is 151 kJ mol^{-1} more stable than cyclohexatriene (Figure 1.1). This is known as the **resonance energy** of ben-

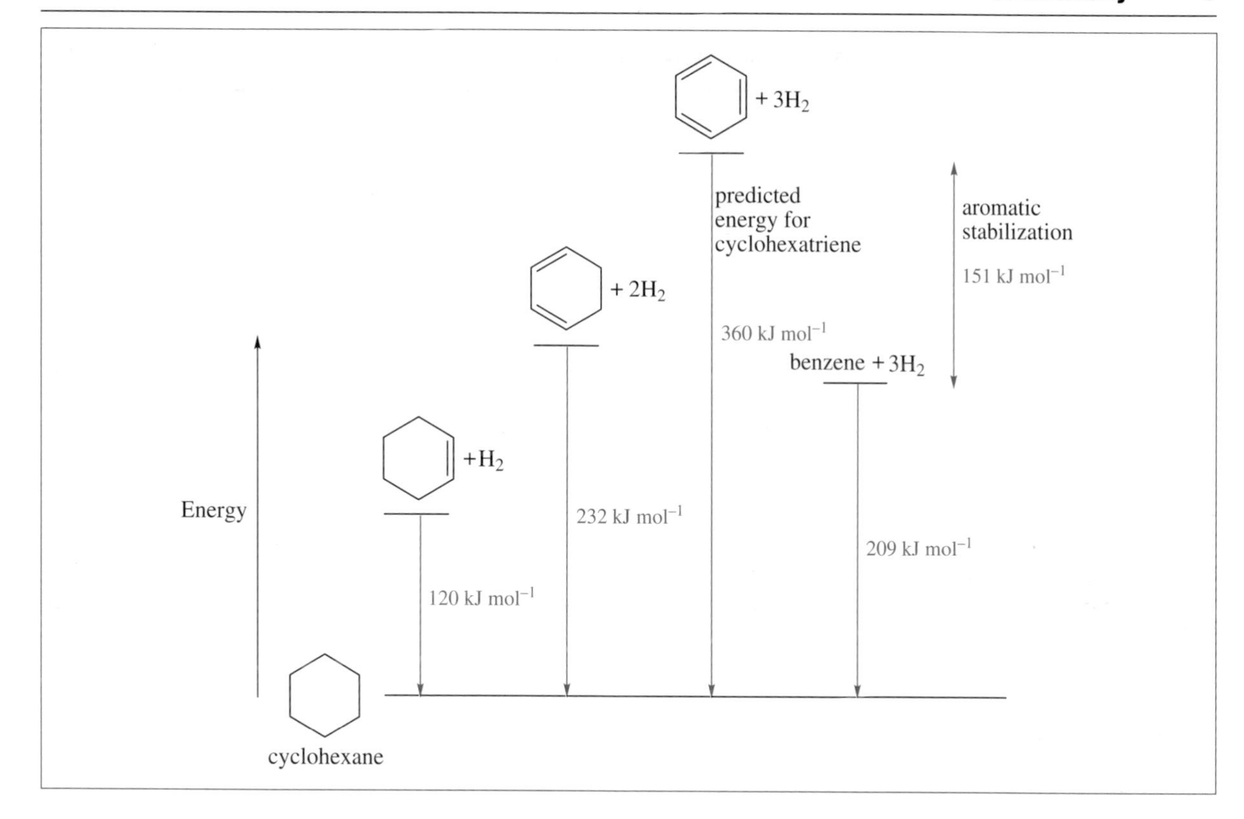

Figure 1.1 Hydrogenation of cyclohexene, cyclohexadiene and benzene

zene or its **aromatic stabilization**. This stabilizing feature dominates the chemistry of benzene and its derivatives.

1.3.1 Valence Bond Theory of Aromaticity

X-ray crystallographic analysis indicated that benzene is a planar, regular hexagon in which all the carbon–carbon bond lengths are 139 pm, intermediate between the single C–C bond in ethane (154 pm) and the C=C bond in ethene (134 pm), and therefore all have some double bond character. Thus the representation of benzene by one Kekulé structure is unsatisfactory. The picture of benzene according to valence bond theory is a resonance hybrid of the two Kekulé or canonical forms **4** and **9**, conventionally shown as in Figure 1.2, and so each carbon–carbon bond apparently has a bond order of 1.5.

A single bond has a bond order of 1 and a double bond a bond order of 2.

Figure 1.2

4 **9** **10** **5** **11** **12**

Kekulé structures

Dewar structures

A double-headed arrow, ↔, is used to connect resonance structures.

It is emphasized that the circle inside a ring represents *exclusively* six π-electrons.

The three Dewar structures **5**, **11** and **12** (Dewar benzene) are also considered to contribute to the resonance hybrid (according to valence bond theory, approximately 20% in total) and to the extra stability. Dewar benzene has now been prepared. It is a bent, non-planar molecule and is not aromatic. It gradually reverts to benzene at room temperature. The Ladenburg structure, prismane (**6**), is an explosive liquid. Dewar benzene and prismane are valence isomers of benzene.

Although the canonical forms for benzene are imaginary and do not exist, the structure of benzene will be represented by one of the Kekulé structures throughout this book. This is common practice. A circle within a hexagon as in **10**, symbolic of the π-cloud, is sometimes used to represent benzene.

1.3.2 Molecular Orbital Theory of Benzene

The current understanding of the structure of benzene is based on molecular orbital (MO) theory. The six carbon atoms of benzene are sp^2 hybridized. The three sp^2 hybrid orbitals of each carbon atom, which are arranged at angles of 120°, overlap with those of two other carbon atoms and with the s orbital of a hydrogen atom to form the planar σ-bonded skeleton of the benzene ring. The p orbital associated with each carbon contains one electron and is perpendicular to the plane of the ring.

MO theory tells us that the six parallel p atomic orbitals are combined together to form six MOs, three of which are bonding orbitals and three anti-bonding. Figure 1.3 shows the relative energy levels of these MOs. The six π-electrons occupy the three bonding orbitals, all of lower energy than the uncombined p orbitals; the higher energy anti-bonding MOs are empty.

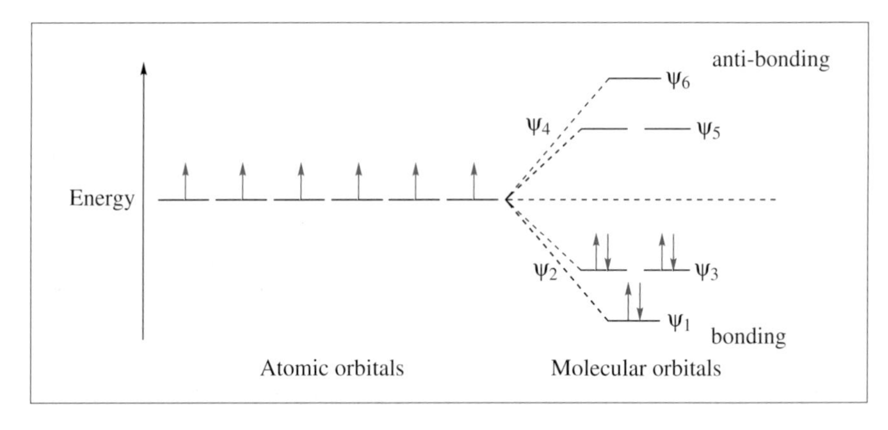

Figure 1.3

The double bond in but-1-ene, $CH_3CH_2CH{=}CH_2$, is fixed between C-1 and C-2; that is, it is localized. However, the double bonds in buta-1,3-diene, $CH_2{=}CH{-}CH{=}CH_2$ are not localized, but are spread over the whole molecule, and are said to be delocalized:

CH_2---CH---CH---CH_2

The same applies to the double bonds in benzene:

This arrangement accounts for the extra stability or aromaticity of benzene. The six overlapping p orbitals can be pictured as forming a **delocalized** π-electron cloud comprising of two rings (think of them as doughnuts!), one above and one below the molecular plane as shown in Figure 1.4. There are no localized C=C bonds as there are in alkenes.

The MOs of benzene are shown pictorially in Figure 1.5. The stability of a MO is related to the number of nodes it possesses; that is to say, the number of times the wave function changes phase (sign) around the ring system. The most stable form has no nodes, when there is a bonding interaction between all six adjacent carbon atoms.

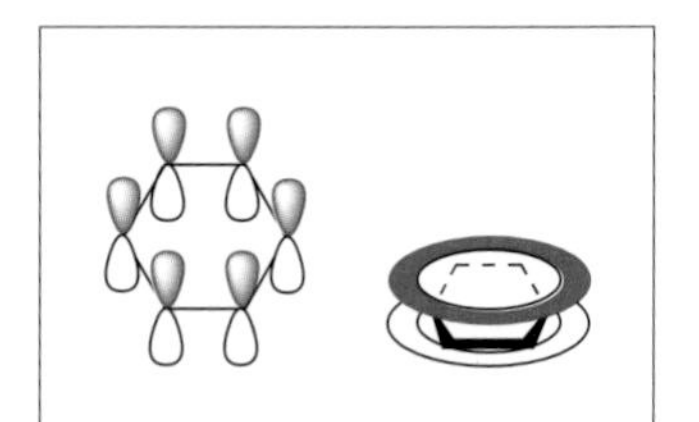

Figure 1.4

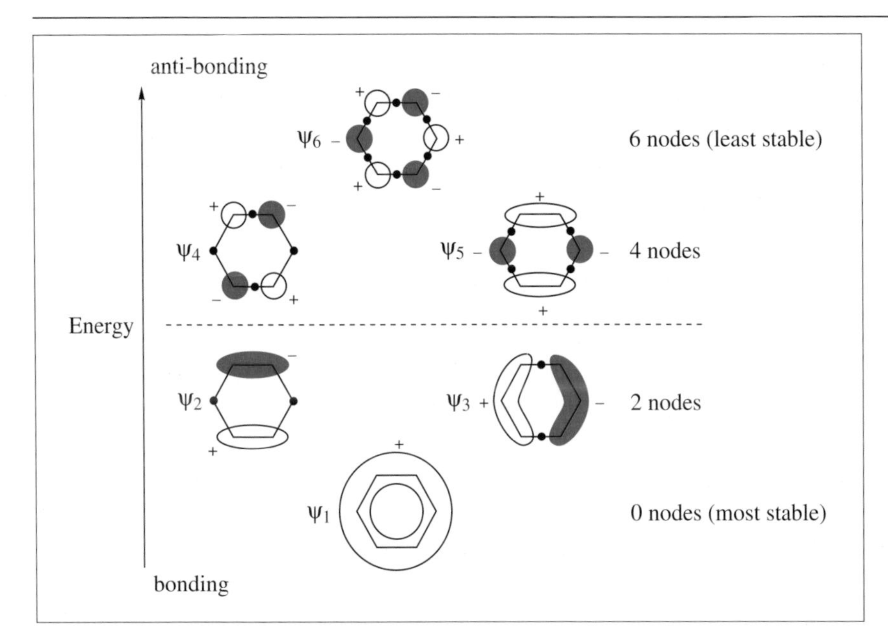

A node is a point where the amplitude of a wave is zero:

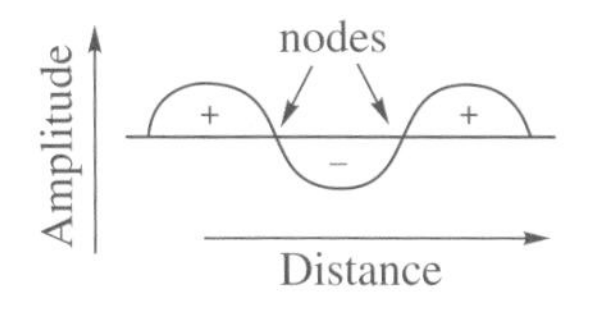

Figure 1.5

1.4 The Hückel Rule

It is important to examine aromaticity in its wider concept at this point. There are many compounds and systems besides benzene that are aromatic. They possess common features in addition to planarity and aromatic stability. MO calculations carried out by Hückel in the 1930s showed that aromatic character is associated with planar cyclic molecules that contained 2, 6, 10, 14 (and so on) π-electrons. This series of numbers is represented by the term $4n + 2$, where n is an integer, and gave rise to Hückel's $4n + 2$ rule that refers to the number of π-electrons in the p-orbital system. In the case of benzene, $n = 1$, and thus the system contains six π-electrons that are distributed in MOs as shown above.

Box 1.1 Hückel's Rule

A planar, cyclic system of unsaturated atoms containing $(4n + 2)$ π-electrons will be aromatic, where n is a positive integer or zero. It will possess extra stabilization. Some slight deviation from planarity is allowed. A similar system but containing $4n$ π-electrons will be **anti-aromatic**. Not only will it lack aromatic stabilization, but the closed loop of π-electrons will result in additional **destabilization** with respect to that anticipated.

These ideas are predicted by calculation and are confirmed by experiment.

This rule is now an important criterion for aromaticity. Those systems that contain $4n$ π-electrons are unstable and are referred to as anti-aromatic compounds.

Degenerate orbitals have equal energies.

The presence of electrons in anti-bonding orbitals is destabilizing. Orbitals that fall on the energy reference line are called nonbonding orbitals; the presence of electrons in these orbitals has no influence on the total bonding.

The reason for the success of the Hückel rule in predicting aromaticity lies in the derivation of the π MOs. For cyclic conjugated molecules, the energy levels of the bonding MOs are always arranged with one lowest-lying MO followed by degenerate pairs of orbitals. The anti-bonding orbitals are arranged inversely, with sets of two degenerate levels and a single highest energy orbital. In the case of benzene, it requires two electrons to fill the first MO and then four electrons to fill each of the n succeeding energy levels, as illustrated in Figure 1.3. A filled set of bonding MOs results in a stable system. This idea is very like that which links the stability of the noble gases to a filled set of atomic orbitals.

Worked Problem 1.1

Q Draw MO diagrams for the π-electron systems of cyclobutadiene and cyclooctatetraene.

A There is a simple system for predicting energy levels of MOs in cyclic systems. A polygon is inscribed within a circle with an apex at the base and horizontal lines are drawn at the points where the polygon touches the circle. The number of points of contact and their position determines the nature of the orbital and the relative energy. A non-bonding orbital results when the polygon touches the circle at a horizontal diameter. Touching below this line indicates bonding orbitals of lower energy; these are filled with electrons first, in accord with Hund's rule. Contact points above the horizontal diametric line correspond to higher energy anti-bonding orbitals. This is illustrated in Figure 1.6 for cyclobutadiene, benzene and cyclooctatetraene, having four, six, and eight carbon atoms, respectively. It can be seen that of these examples only benzene obeys Hückel's $4n + 2$ rule; both cyclobutadiene and cyclooctatetraene apparently have unpaired electrons. However, cyclooctatetraene is not planar, existing in a tub shape, and it behaves as a polyalkene with all the π-electrons paired up.

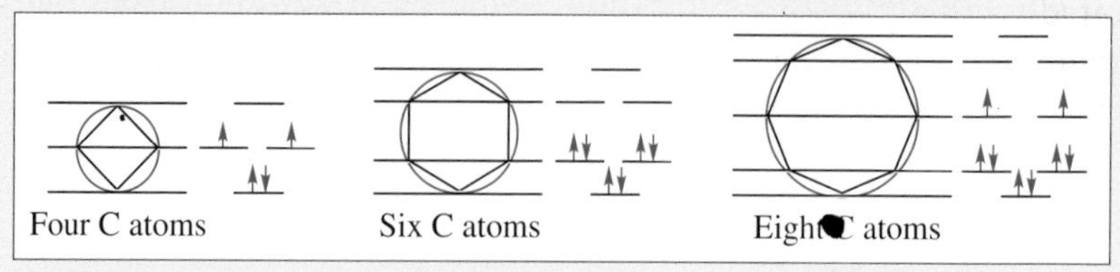

Figure 1.6

This method can also be used for other simple monocyclic systems that can be inscribed in a circle.

Although adherence to the Hückel rule is a valuable test for aromaticity, other properties are also used to assess whether a compound is aromatic or not. One such diagnostic tool is ^{1}H NMR spectroscopy. When exposed to a magnetic field, the π-electron cloud circulates to produce a ring current that generates a local magnetic field (Figure 1.7). This new field boosts the applied magnetic field outside the ring. As a result, the hydrogen atoms are deshielded and resonate at a lower applied field, usually in the range δ 6.5–8.5 ppm. Alkenyl hydrogen atoms are also deshielded, but to a lesser extent and normally resonate in the region δ 4.5–5.5 ppm. The local field inside the ring opposes the applied field and this effect is apparent in the ^{1}H NMR spectra of the annulenes (see p. 11).

The chemical shift of the resonance signal of a proton, δ_H, is measured relative to a standard, tetramethylsilane, $(CH_3)_4Si$, for which δ_H is 0.00 ppm. It reflects the chemical environment of the proton. A similar situation applies to the resonance signal δ_C for a ^{13}C nucleus.

Aromatic compounds also show characteristic infrared and ultraviolet absorption spectra. In the mass spectrum of aromatic substances, peaks corresponding to ions such as $C_6H_5^+$ and $C_6H_6^+$ are often seen. A commonly observed peak occurs at *m/z* 91, corresponding to the stable ion $C_6H_5CH_2^+$. These features are all helpful in assigning aromatic character.

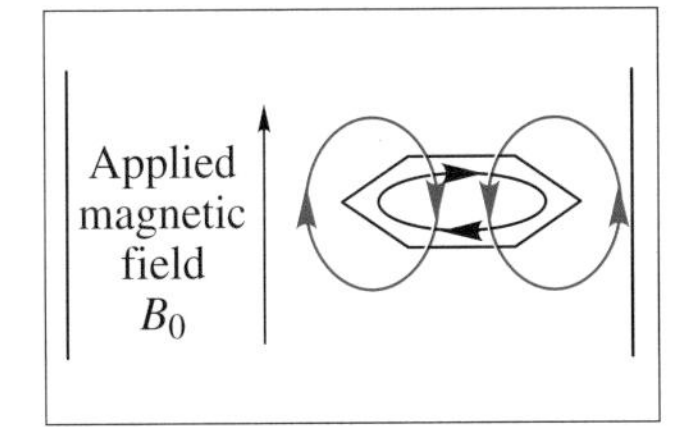

Figure 1.7

1.4.1 2π-Electron Systems

Aromatic systems that obey Hückel's $4n + 2$ rule where $n = 0$ and so possess two π-electrons do exist and are indeed stable. The smallest possible ring is three membered and the derived unsaturated structure is cyclopropene. The theoretical loss of a hydride ion from this molecule leads to the cyclopropenyl cation, which contains two π-electrons distributed over the three carbon atoms of the planar cyclic system (Figure 1.8).

A circle with a charge inside it and inscribed within a ring structure is used to denote a $(4n + 2)$ π-electron system.

H H
sp^3
−H$^-$
sp^2
non-planar
not aromatic
2π-electron system

H
sp^2
planar
"aromatic"
2π-electron system

Figure 1.8

This cationic species and a number of its derivatives have been prepared and they are quite stable, despite the strain associated with the internal bond angles of only 60°. For example, the reaction of hydrogen bromide with diphenylcyclopropenone, which is itself a stable compound with aromatic character, gives the diphenylcyclopropenium salt (Scheme 1.1).

Ph, Ph, O — HBr → Ph, Ph, OH Br$^-$

Scheme 1.1

Examination of the cyclobutadiene system indicates that it possesses four π-electrons and is thus an unstable $4n$ system. Cyclobutadiene itself only exists at very low temperatures, though some of its derivatives are stable to some extent at room temperature. Cyclobutadiene is a rectangular diene. Loss of two electrons through the departure of two chloride ions from the 3,4-dichlorocyclobutene derivative creates a 2π-electron aromatic system, the square, stable cyclobutenyl dication (Scheme 1.2).

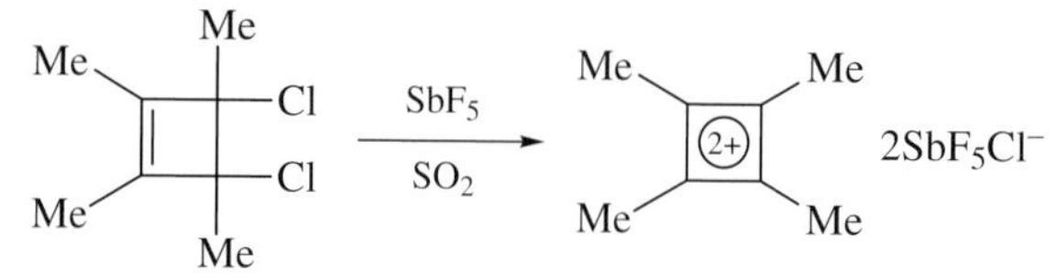

Me, Me, Me, Me, Cl, Cl; SbF_5 / SO_2 → Me, Me, Me, Me (2+) $2SbF_5Cl^-$

Scheme 1.2

1.4.2 6π-Electron Systems

We have seen that benzene fits into this category, but there are a number of other stable aromatic systems that contain six π-electrons.

Cyclopentadiene is surprisingly acidic (pK_a *ca.* 16) for a hydrocarbon. This property arises because the cyclopentadienyl anion, generated by abstraction of a proton by a base such as sodium ethoxide (Scheme 1.3), has a delocalized aromatic set of six π-electrons.

A curly (or curved) arrow, ⌒, is used to show the movement of an electron pair.

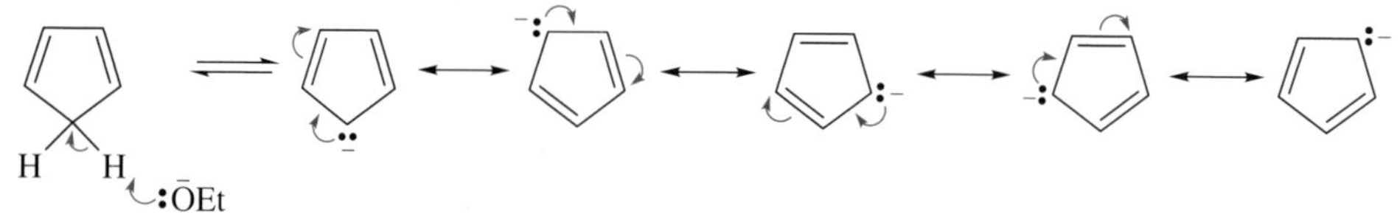

H H :ŌEt

Scheme 1.3

The cyclopentadienyl anion **13** is an efficiently **resonance-stabilized anion** in which all the carbon–carbon bond lengths are equal (Figure 1.9). It forms stable compounds, of which ferrocene (**14**) is an example, which undergo aromatic substitution reactions such as sulfonation and acetylation.

In contrast, it is the **resonance-stabilized cation** derived from cycloheptatriene that possesses the aromatic sextet of π-electrons. Tropylium bromide is formed by the addition of bromine to cycloheptatriene and

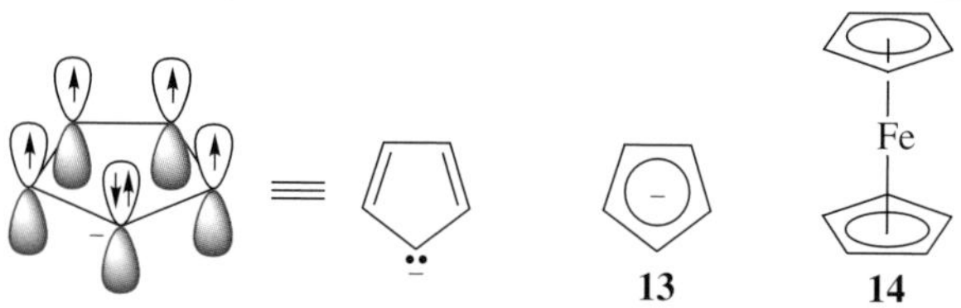

Fe 13 14

Figure 1.9

then loss of hydrogen bromide by heating. It can also be generated directly from cycloheptatriene by hydride ion abstraction using triphenylcarbenium perchlorate (Scheme 1.4). In the tropylium ion **15**, the bond lengths are equal and all seven carbon atoms share the positive charge (Figure 1.10).

Unlike cyclopentadiene, cycloheptatriene is not an acidic hydrocarbon; its pK_a is about 36. If a proton could be abstracted from cycloheptatriene, the resulting anion would have eight π-electrons and would be an unstable, anti-aromatic system.

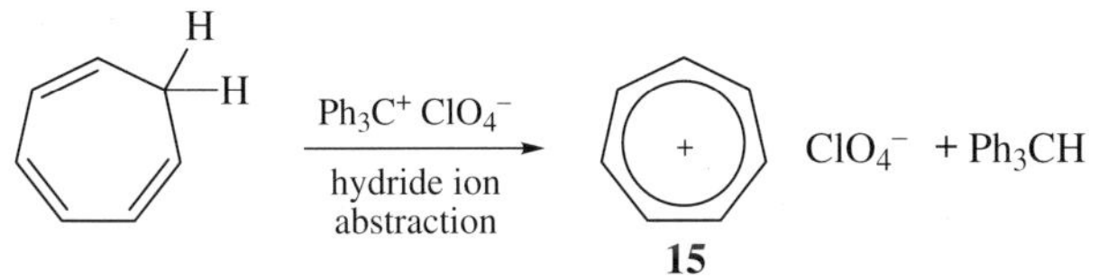

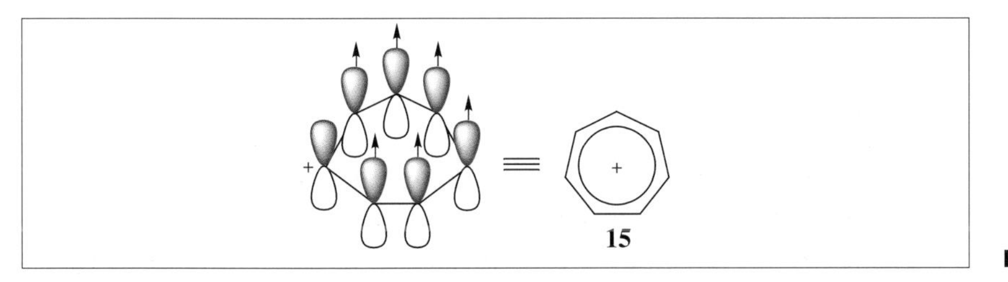

Scheme 1.4

Figure 1.10

Worked Problem 1.2

Q Write an equation to show the formation of the cycloheptatrienyl cation by addition of bromine to the hydrocarbon. Use curly arrows to illustrate the stabilization of the cation.

A Bromine adds to one double bond of cycloheptatriene in exactly the same way as it does to ethene (although conjugate 1,4- and 1,6-addition may be the preferred mode of reaction). Dehydrobromination occurs on heating. The driving force for the loss of bromide ion is the extra stability of the resulting cation. The positive charge is delocalized over the whole system (Scheme 1.5).

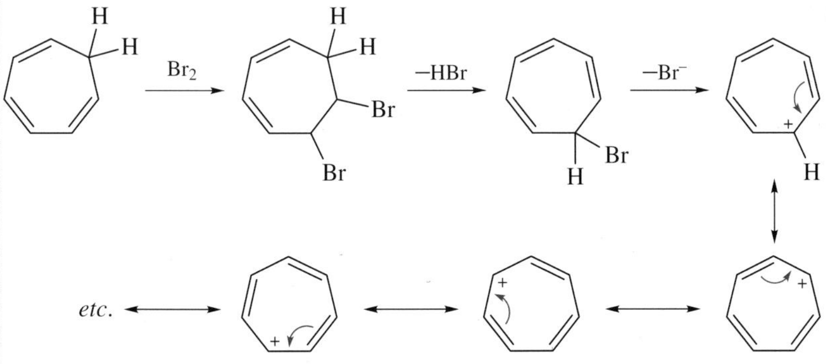

Scheme 1.5

A dipole moment is associated with a permanent uneven sharing of electron density in a molecule. It is a vector quantity that depends on the size of the partial charges in the molecule and the distance between them, and is measured in units of Debye (D). For example, chlorobenzene has a dipole moment of 1.75 D, oriented towards the chlorine. In 1,4-dichlorobenzene, the individual moments directed towards the chlorine atoms are equal and opposed to each other. Consequently, the dipole moment of this molecule is zero.

Azulene (**16**) is a stable, blue solid hydrocarbon that undergoes typical electrophilic aromatic substitution reactions. It may be regarded as a combination of **13** and **15**; in keeping with this it has a dipole moment of 0.8 D (Figure 1.11). The fusion bond linking the two rings is longer (150 pm) than the other bonds (139–140 pm), indicating that azulene is a peripherally conjugated system.

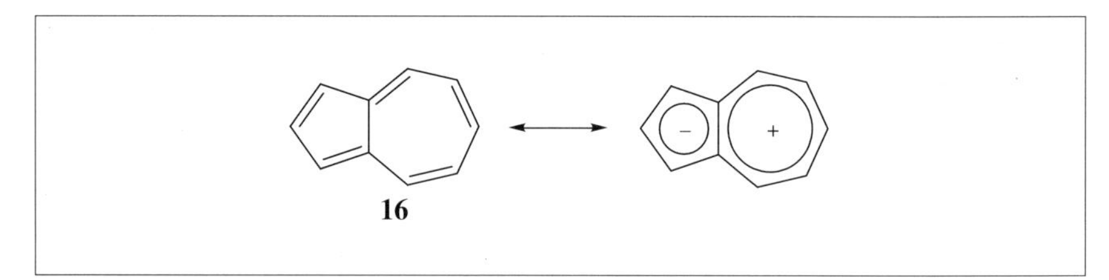

Figure 1.11

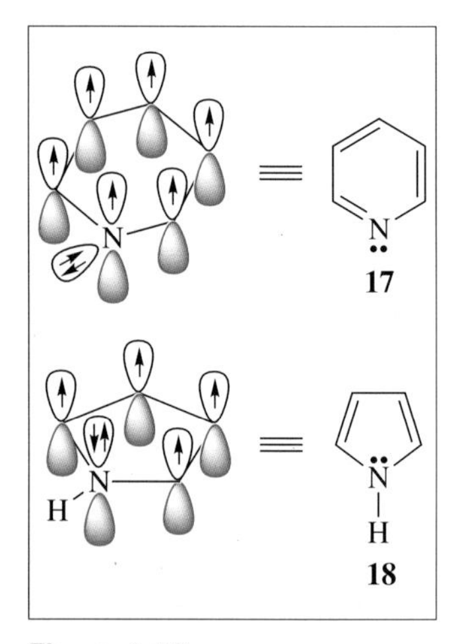

Figure 1.12

Some heterocyclic compounds possess aromatic character. One such important compound is pyridine (**17**), in which one of the CH units of benzene has been replaced by a nitrogen atom (Figure 1.12). Although the chemistry of pyridine shows several important differences from benzene, it also has some common characteristics. The five carbon atoms and the nitrogen atom each provide one electron for the π-cloud, thereby conferring aromaticity on pyridine according to Hückel's rule. Notice that the nitrogen retains a lone pair of electrons in an sp^2 orbital directed away from the ring; this accounts for the basic properties of pyridine.

Similarly, the five-membered heterocycle pyrrole (**18**) is aromatic, although this molecule obeys Hückel's rule only because the nitrogen atom contributes two electrons to the π-cloud. In this respect, pyrrole is analogous to the cyclopentadienyl anion. As a consequence, the nitrogen atom does not retain a lone pair of electrons and pyrrole is not basic.

Addition of two electrons to cyclooctatetraene would lead to a 10π-electron system, the cyclooctatetraenyl dianion. Cyclooctatetraene, which is non-planar, reacts with potassium in tetrahydrofuran to give dipotassium cyclooctatetraenide, which is planar. The C–C bond lengths are all the same (141 pm). At low temperatures, a dication is formed when cyclooctatetraene reacts with SbF_5. This also has some aromatic characteristics in keeping with a 6π-electron system.

1.4.3 10π-, 14π- and 18π-Electron Systems

The most important 10π carbocyclic system is naphthalene (**19**) in which two benzene rings are fused together. The fused systems anthracene (**20**) and phenanthrene (**21**) obey Hückel's rule, where n = 3, and have 14π-electrons. All three compounds are typically aromatic and their chemistry is similar to that of benzene, as discussed in Chapter 12.

19 20 21

In 1962, Sondheimer prepared a series of conjugated monocyclic polyenes called annulenes, with the specific purpose of testing Hückel's rule. Amongst the annulenes prepared, compound **22** with 14 and compound **23** with 18 carbon atoms, that is *n* =3 and *n* = 4, respectively, have the magnetic properties required for aromatic character, but behave chemically like conjugated alkenes. In [18]annulene (**23**), the hydrogen atoms on the outside of the ring resonate in the aromatic region at δ 9.3 ppm. However, the inner protons lie in the region where the induced field associated with the ring current opposes the applied field. They are therefore shielded and so resonate upfield at δ –3.0 ppm.

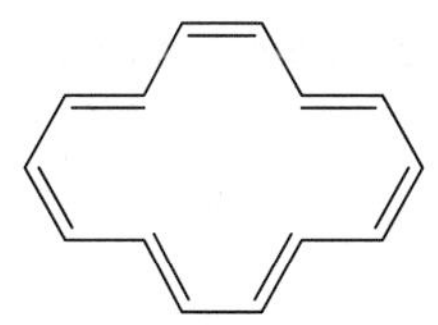

22

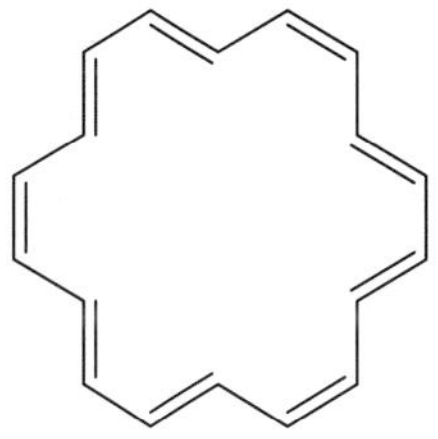

23

1.5 Nomenclature

The remainder of this book will be devoted to the synthesis and reactions of a range of aromatic compounds. It is important that you understand the naming of these compounds. The use of trivial names is widespread, particularly in the chemical industry; although some of the older names have disappeared from use, many persist and are allowed in the IUPAC system. Some of these are presented in Figure 1.13.

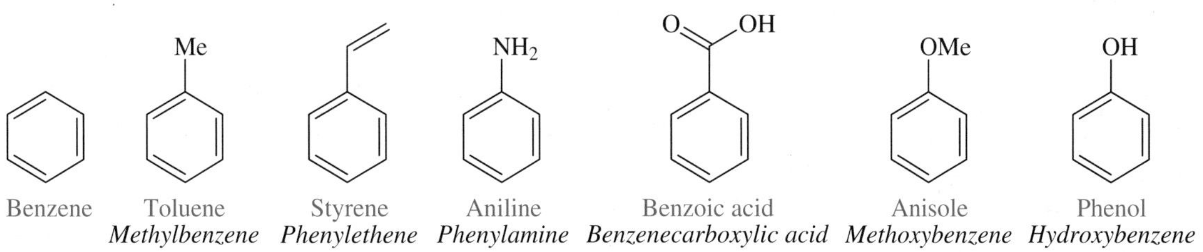

Me
NH2
O
OH
OMe
OH
Benzene
Toluene
Methylbenzene
Styrene
Phenylethene
Aniline
Phenylamine
Benzoic acid
Benzenecarboxylic acid
Anisole
Methoxybenzene
Phenol
Hydroxybenzene

Figure 1.13

Monosubstituted compounds are commonly named as in aliphatic chemistry, with the substituents appearing as a prefix to the parent name benzene; bromobenzene, chlorobenzene and nitrobenzene are examples (Figure 1.14).

Br
Cl
NO2
Bromobenzene
Chlorobenzene
Nitrobenzene

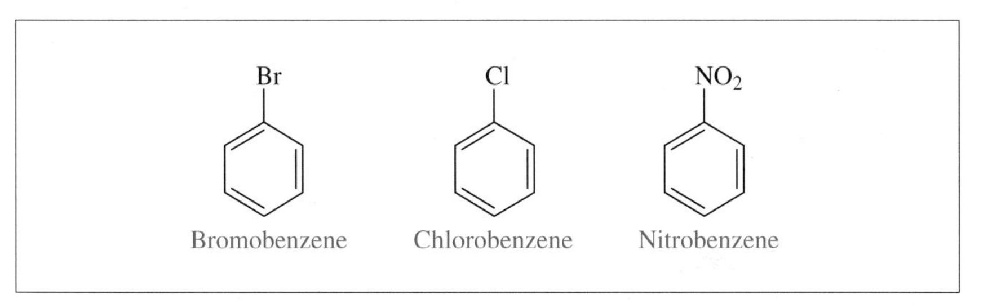

Figure 1.14

There are two acceptable ways of naming the three positional isomers that are possible for disubstituted benzene rings. The substituent

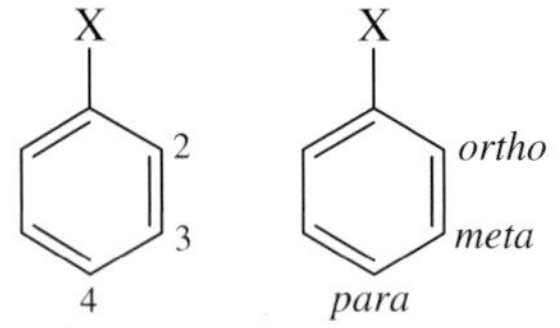
X
2
3
4
X
ortho
meta
para

24 25

positions 1,2-, 1,3- and 1,4- are sometimes replaced by the terms *ortho*-, *meta*- and *para*- (abbreviated to *o*-, *m*- and *p*-, respectively) (see **24** and **25**). You are advised to become familiar with both systems so that you can use them interchangeably.

In multiply substituted compounds, the groups are numbered so that the lowest possible numbers are used. The substituents are then listed in alphabetical order with their appropriate numbers. Examples are given in Figure 1.15, which also introduces further trivial names.

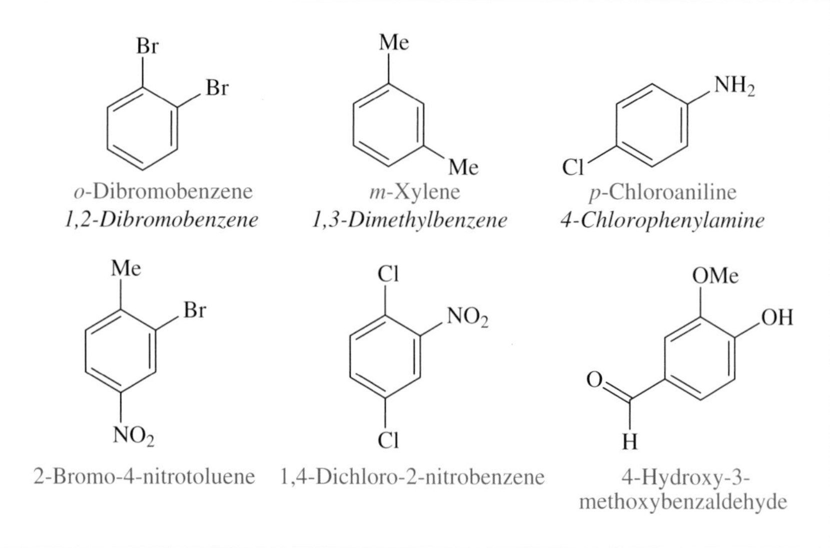
Br
Br
o-Dibromobenzene
1,2-Dibromobenzene
Me
Me
m-Xylene
1,3-Dimethylbenzene
NH_2
Cl
p-Chloroaniline
4-Chlorophenylamine
Me
Br
NO_2
2-Bromo-4-nitrotoluene
Cl
NO_2
Cl
1,4-Dichloro-2-nitrobenzene
OMe
OH
O
H
4-Hydroxy-3-methoxybenzaldehyde

Figure 1.15

There are occasions when the benzene ring is named as a substituent and in these cases the name for C_6H_5- is phenyl, abbreviated to Ph. The name for $C_6H_5CH_2$- is benzyl or Bn, whilst the benzoyl substituent is C_6H_5CO- or Bz. These substituents can also be named systematically as shown in Figure 1.16.

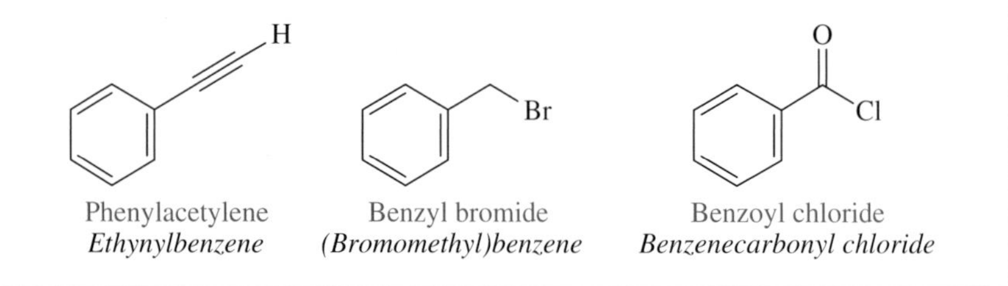
H
Phenylacetylene
Ethynylbenzene
Br
Benzyl bromide
(Bromomethyl)benzene
O
Cl
Benzoyl chloride
Benzenecarbonyl chloride

Figure 1.16

Summary of Key Points

1. Benzene is an unusually stable molecule. This aromatic stabilization is associated with its 6π-electron system.

2. Other molecules besides benzenoid compounds also show increased stability.

3. Aromatic stability is found in planar, cyclic conjugated systems that satisfy the Hückel rule in possessing $(4n + 2)$ π-electrons.

4. $(4n + 2)$ π-electrons corresponds to a completely filled set of bonding molecular orbitals.

5. Ionic species that comply with all the requirements for aromaticity are best regarded as resonance-stabilized ions.

6. Protons in an aromatic environment usually resonate at δ 6.5–8.5 ppm.

7. Cyclic, planar molecules with $4n$ π-electrons are particularly unstable and are said to be anti-aromatic.

8. Substitution reactions, which are the most important type shown by aromatic compounds, conserve aromatic stability.

9. Addition reactions, which are less common, destroy the aromatic system.

Problems

1.1. Draw the MO diagrams of the π-electron system for the following species: the cyclopropenyl radical, the cyclopropenyl cation, the cyclobutadienyl cation, the cyclopentadienyl anion, the tropylium cation, the cyclooctatetraenyl dication.

1.2. Cyclopropenones are described as having aromatic character. How would you account for this, given that the ring contains three π-electrons?

1.3. Which of the following compounds would you expect to be aromatic:

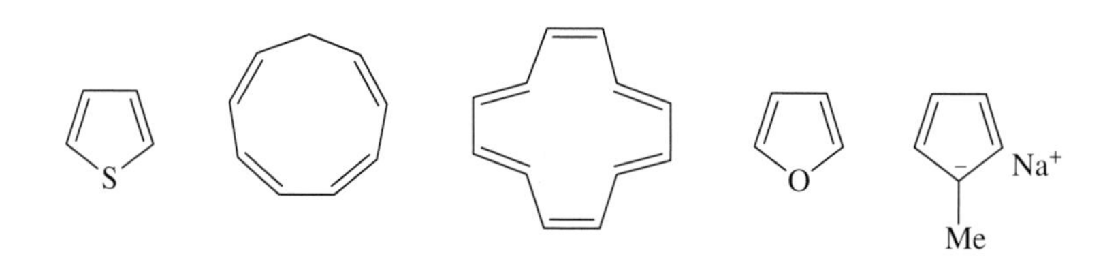

1.4. Give the systematic names for the following compounds:

NH_2 Cl NO_2 (a)

HO_3S Br (b)

CO_2H H_2N NO_2 (c)

OH Cl Cl (d)

Cl CO_2H (e)

1.5. Draw structures for the following compounds: (a) 4-chloro-3-nitrobenzoic acid; (b) *m*-nitroethylbenzene; (c) 2,6-dibromo-4-nitroaniline; (d) *p*-chlorobenzenesulfonic acid; (e) *o*-nitrobenzaldehyde

2

Aromatic Substitution

Aims

By the end of this chapter you should understand:

- The mechanism of electrophilic aromatic substitution
- How a variety of substituents can be substituted into the benzene ring
- The effects of substituents on orientation and reactivity
- The mechanism of nucleophilic substitution in aromatic systems

2.1 Introduction

In Chapter 1 it was stated that the principal reaction of benzene and its derivatives is **substitution** rather than addition. Indeed, electrophilic substitution in aromatic systems is one of the most important reactions in chemistry and has many commercial applications.

The π-electron cloud above and below the plane of the benzene ring is a source of electron density and confers nucleophilic properties on the system. Thus, reagents that are deficient in electron density, **electrophiles**, are likely to attack, whilst electron-rich nucleophiles should be repelled and therefore be unlikely to react. Furthermore, in electrophilic substitution the leaving group is a proton, H^+, but in nucleophilic substitution it is a hydride ion, H^-; the former process is energetically more favourable. In fact, **nucleophilic aromatic substitution** is not common, but it does occur in certain circumstances.

The carbocation generated by the addition of an electrophile to an alkene is destroyed in the second step by the addition of a nucleophilic species:

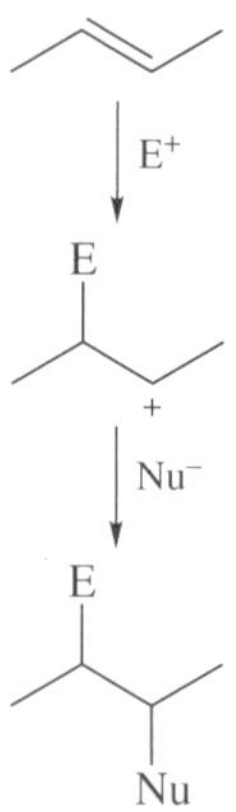

E+

E

+

Nu−

E

Nu

Both carbon atoms become sp^3 hybridized and the double bond is lost. A similar second step in aromatic molecules would result in destruction of the resonance-stabilized system and therefore does not occur.

2.2 Electrophilic Aromatic Substitution (S_EAr)

In simple terms, electrophilic aromatic substitution proceeds in two steps. Initially, the electrophile E^+ adds to a carbon atom of the benzene ring in the same manner in which it would react with an alkene, but here the π-electron cloud is disrupted in the process. However, in the second step the resultant carbocation eliminates a proton to regenerate the aromatic system (Scheme 2.1). The combined processes of addition and elimination result in overall substitution.

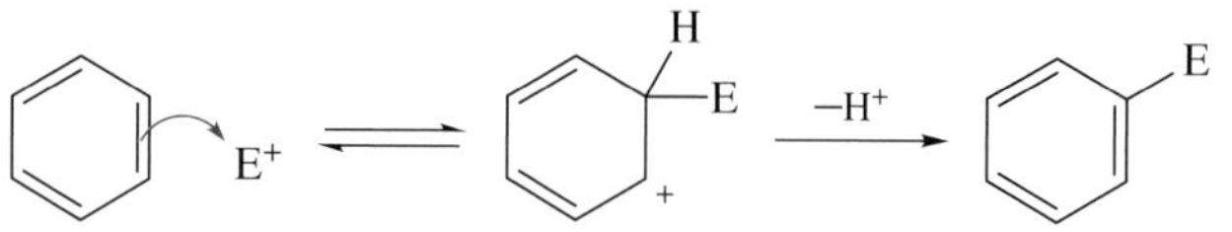

H E+ E −H+ + E

Scheme 2.1

The hybridization state of the carbon atom that is attacked changes from sp^2 to sp^3 and the planar aromatic system is destroyed. An unstable **carbocation** is simultaneously produced and so it is clear that this step is energetically unfavourable. It is therefore the slower step of the sequence.

However, the intermediate carbocation is stabilized by resonance, with the positive charge shared formally by three carbon atoms of the benzene ring (Scheme 2.2). The resonance hybrid structure **1** indicates the delocalization of the charge. The carbocation is also referred to as a **σ-complex** or **Wheland intermediate**.

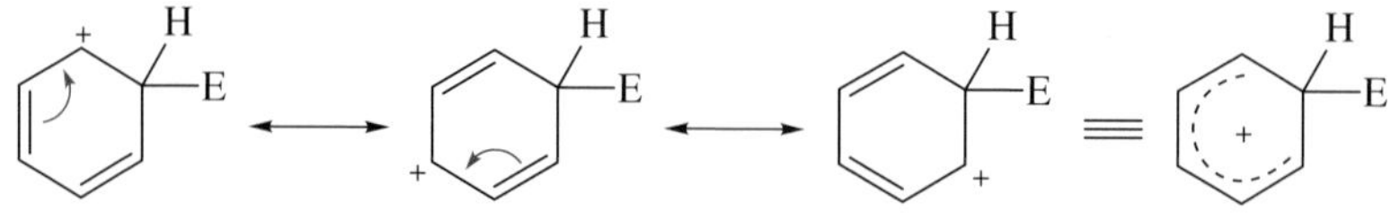

+ H E H E + H E + H E + 1

In the second step, a proton is abstracted by a basic species present in the reaction mixture. The attacked carbon atom reverts to sp^2 hybridization and planarity and aromaticity are restored. This fast step is energetically favourable and is regarded as the driving force for the overall process. The product is a substituted benzene derivative.

The energy changes that occur during the course of the reaction are related to the structural changes in the reaction profile shown in Figure 2.1. It should be noted that each step proceeds through a high-energy transition state in which partial bonds attach the electrophile and the proton to the ring and the π-cloud is incomplete.

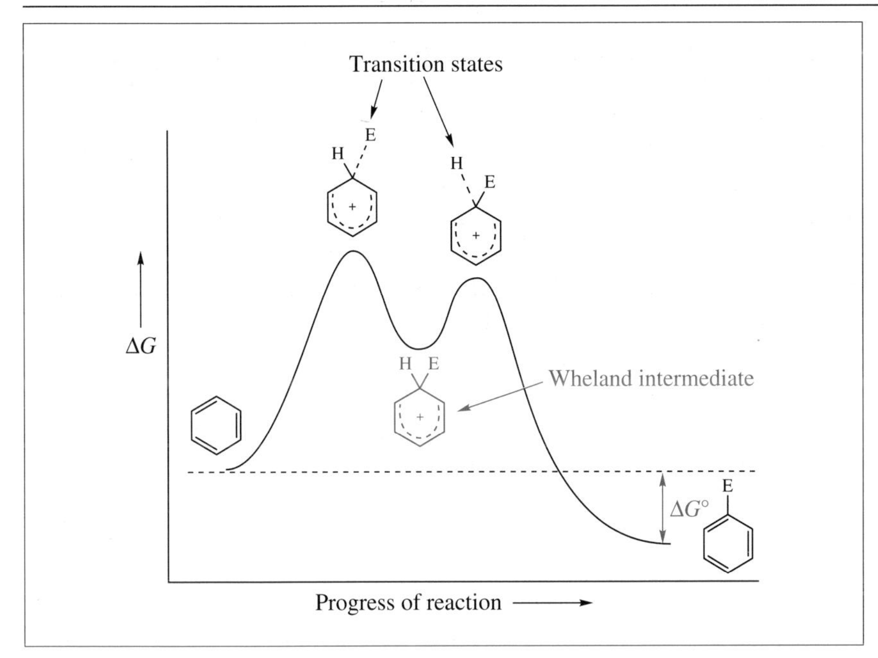

Figure 2.1 Energy profile for electrophilic attack on benzene

In some instances, the Wheland intermediates can be isolated. For example, when hexamethylbenzene (**2**) is treated with the nitrating agent $NO_2^+BF_4^-$, the product **3** cannot undergo proton loss and can be isolated at −70 °C.

Most examples of electrophilic aromatic substitution proceed by this sequence of events:

- Generation of an electrophile
- The electrophile attacks the π-cloud of electrons of the aromatic ring
- The resulting carbocation is stabilized by resonance
- A proton is abstracted from the carbocation, regenerating the π-cloud
- A substituted aromatic compound is formed

In the following sections, various examples are reviewed, highlighting the source of the electrophile and any variations in mechanistic detail. Further discussion of the reactions and the products will be found in Chapters 4–9, which deal with the chemistry of functionalized derivatives of benzene.

2.2.1 Nitration of Benzene

Benzene cannot be nitrated using nitric acid alone, which lacks a strong electrophilic centre, but it is readily achieved using a mixture of concentrated nitric acid and concentrated sulfuric acid, the so-called "mixed acid". The product is nitrobenzene. The interaction of nitric acid and sulfuric acid produces the electrophile, the nitronium ion NO_2^+, according to Scheme 2.3. The sulfuric acid is also the source of the base HSO_4^- that removes the proton in the second step.

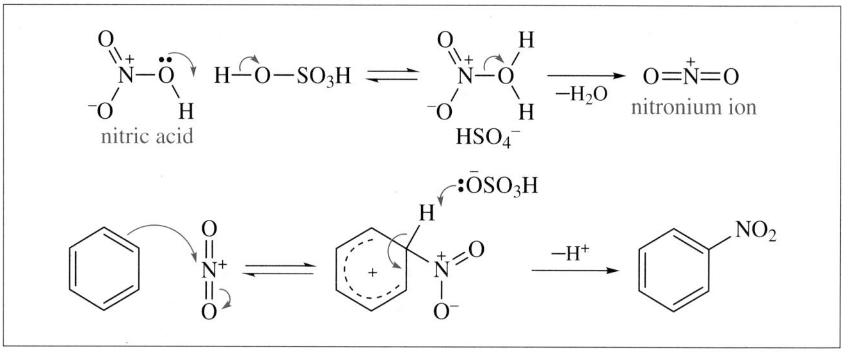

Scheme 2.3

2.2.2 Halogenation of Benzene

A Lewis acid is a species which accepts an electron pair. A Lewis base donates an electron pair.

Halogen molecules are not strong electrophiles and, fluorine excepted, do not react with benzene. However, in the presence of a Lewis acid, reaction occurs readily. The role of the catalyst is to accept a lone pair of electrons from the halogen molecule, which then becomes electron deficient at one of the halogen atoms. The actual electrophile is probably the complex formed from the halogen and the catalyst, rather than a halonium ion, *e.g.* Cl^+ or Br^+. Bromination of benzene serves as a good example of halogenation (Scheme 2.4).

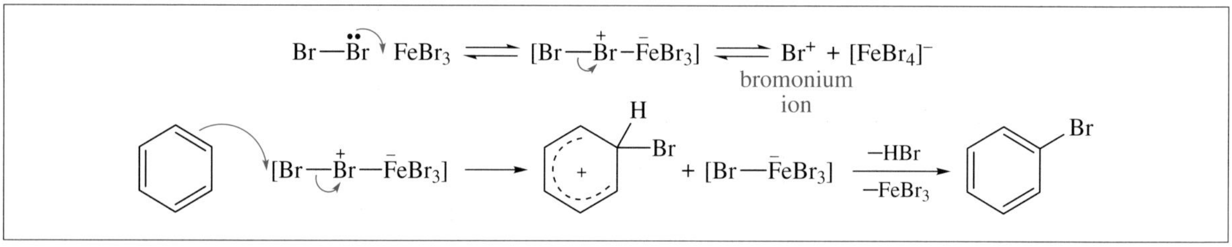

Scheme 2.4

2.2.3 Friedel–Crafts Alkylation

A new C–C bond is formed in Friedel–Crafts alkylation and acylation reactions.

Alkyl halides require a Lewis acid catalyst to accentuate the polarization and create a more powerful electrophile. There is not enough positive character on the carbon atom in alkyl halides for them to react with benzene; the catalyst increases the positive character. Aluminium

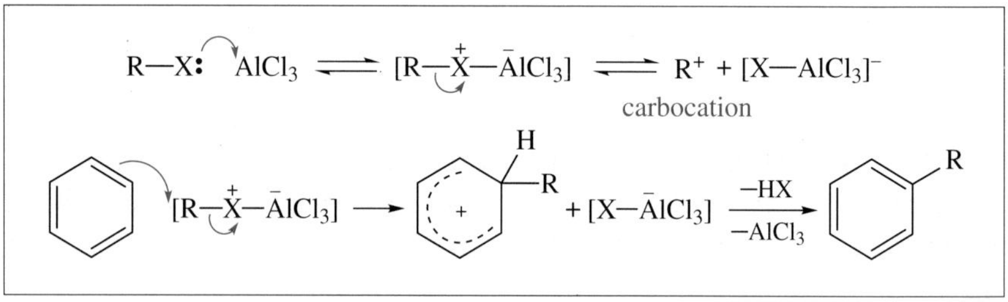

Scheme 2.5

chloride is commonly used as the Lewis acid, accepting a pair of electrons from the halogen atom (Scheme 2.5). The electrophile may be a carbocation or perhaps more likely the complex shown. An alkylbenzene is produced.

2.2.4 Friedel–Crafts Acylation

Acylation can be achieved using either acyl halides or acid anhydrides. The product is a ketone. Acyl halides are more reactive than the anhydrides, but still require a Lewis acid catalyst to promote the reaction (Scheme 2.6). The attacking species is the resonance-stabilized acylium ion or the complex.

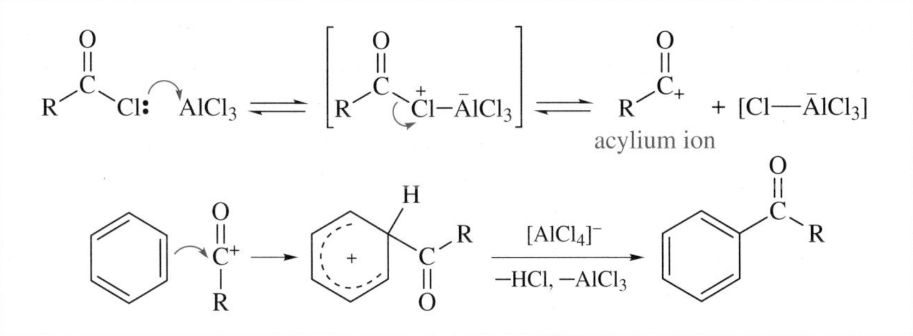

acylium ion

$[AlCl_4]^-$

$-HCl, -AlCl_3$

Scheme 2.6

2.2.5 Sulfonation of Benzene

Benzene itself is not attacked by concentrated sulfuric acid, but is readily converted to benzenesulfonic acid by fuming sulfuric acid. This is a solution of sulfur trioxide in concentrated sulfuric acid, and is known as oleum. Note here that the attacking electrophile is a neutral species and that the electron-deficient sulfur atom of SO_3 is the electrophilic centre (Scheme 2.7).

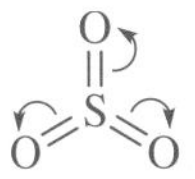

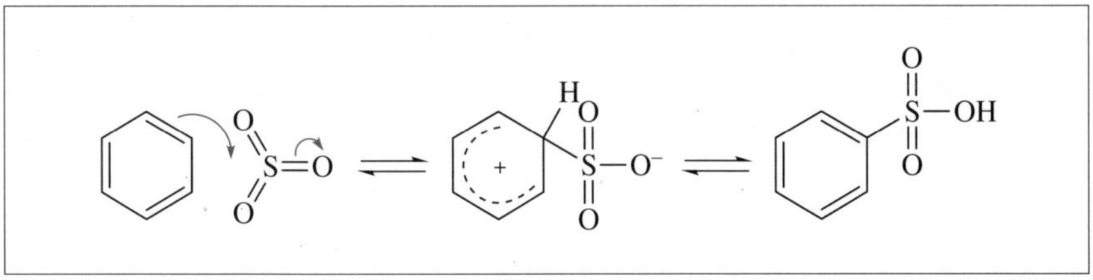

Scheme 2.7

Sulfonation differs from the other examples which have been discussed in that it can readily be reversed. Heating benzenesulfonic acid with dilute sulfuric acid or water converts it back to benzene.

Because sulfonation is reversible, a sulfonic acid group is sometimes introduced to occupy a position on the ring, temporarily protecting that position from attack by another electrophile. Sulfonation is also a useful way of directing further substitution to a specific position.

2.2.6 Protonation

Although benzene is a very weak base, it is protonated in concentrated sulfuric acid to a very slight extent. This reaction can be detected if the protonating mixture contains deuterium or tritium, the isotopes of hydrogen, since isotope exchange takes place. Some deuteriated benzene is produced when benzene is treated with D_2SO_4, and this can be detected by mass spectrometry and NMR spectroscopy. The more deuterium there is in the protonation mixture, the more exchange occurs. Notice that the regeneration of the aromatic system occurs by elimination of a proton (Scheme 2.8).

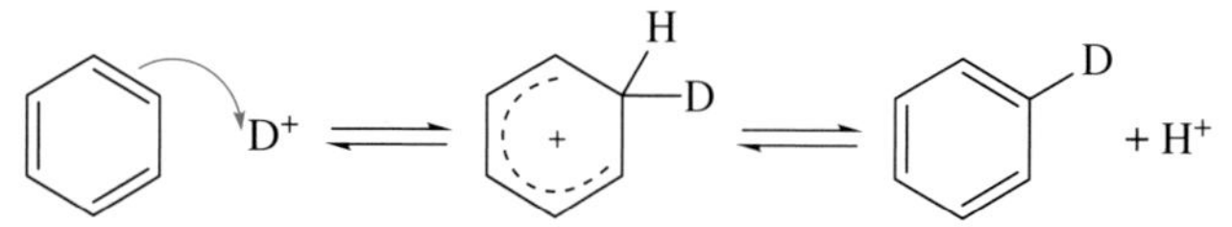

D^+ ⇌ H, D, + ⇌ D, $+ H^+$

Scheme 2.8

Exercise 2.1 Electrophilic Substitution

Copy out this revision table about electrophilic aromatic substitution reactions of benzene and fill in the relevant electrophilic species and draw structures for the products

Reaction	*Reagents*	*Electrophile*	*Product*
Nitration	HNO_3, H_2SO_4		
Halogenation	Cl_2, $AlCl_3$		
Friedel–Crafts alkylation	Me_2CHOH, conc. H_2SO_4		
Friedel–Crafts acylation	AcCl, $AlCl_3$		
Sulfonation	H_2SO_4, SO_3		

Ac is an abbreviation for the acetyl group, CH_3CO.

2.3 Reactivity and Orientation in Electrophilic Aromatic Substitution

How do derivatives of benzene behave towards electrophilic attack? Two experimental observations illustrate that the behaviour is quite varied. The rate of nitration of toluene is appreciably faster than that of benzene and produces a mixture of 2- and 4-nitrotoluenes. On the other hand, the nitration of nitrobenzene is more difficult than that of benzene and gives just one product, 1,3-dinitrobenzene (Scheme 2.9).

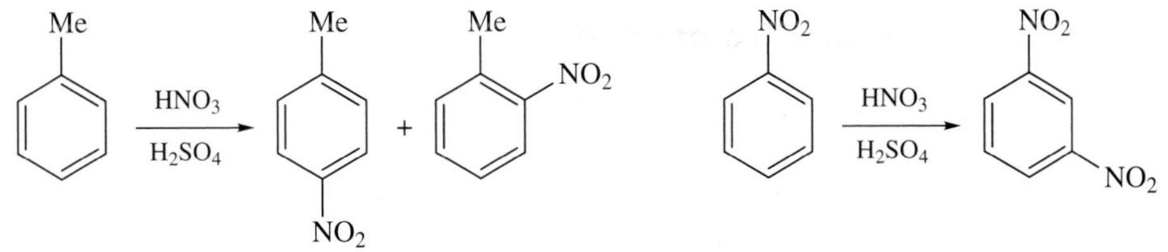

Scheme 2.9

A substituent in a benzene ring therefore influences the course of electrophilic substitution in two ways:

- It affects the reactivity of the molecule
- It controls the orientation of attack, *i.e.* which isomer is formed

It is important to understand why this should happen. In the above examples, the two substituents, the methyl group and the nitro group, exhibit different electronic behaviour. The methyl group is an **electron donor** and so increases the electron density of the ring. The nitro group is an **electron acceptor** and withdraws electron density from the ring.

It is these properties that influence the course of the reactions of aromatic compounds with electrophiles. An electron-releasing group increases the electron density of the benzene ring, promoting electrophilic attack. Such substituents are known as **activating groups**. An electron-withdrawing group is **deactivating** and reduces the electron density of the ring, making attack by the electron-deficient reagent more difficult.

Both types of substituents affect the electron density at all positions of the ring, but exert their greatest effects at the *ortho* and *para* positions, making these sites the most electron rich in the case of donor groups and most electron deficient when electron-withdrawing groups are present. Donor groups therefore direct attack of the electrophile to the *ortho* and *para* positions and are known as ***ortho/para* directors**. Conversely, aromatic compounds containing electron acceptor groups are attacked at the *meta* position since this is the least electron-deficient site. Such groups are called ***meta* directors**. Not all substituents fit exactly into this picture: halogens are deactivating but direct attack to the *ortho* and *para* positions.

- Electron-donating substituents activate the benzene ring to electrophilic attack, which results in the formation of the *ortho*- and *para*-disubstituted benzene derivatives.
- Electron-withdrawing substituents deactivate the ring to attack by electrophiles, which occurs at the *meta* position.

Substituents exert their influence on a molecule through either the σ-bonds or the π-bonding system, in other words by inductive and mesomeric (resonance) effects, respectively (see below). The interaction influences both the electron density at the various ring positions and the stability of the intermediate carbocation. The outcome can be understood by superimposing the electronic effects of the substituents on the slow, rate-determining step of the general mechanism for electrophilic aromatic substitution discussed above.

In a σ-bond between two atoms of differing electronegativities there is an unequal sharing of the electron pair, with the electrons being attracted towards the more electronegative atom. This causes a permanent polarization of the molecule. This influence of an atom or group on the distribution of the electron pair is called the **inductive effect**. Inductive effects rapidly die away along a saturated carbon chain (see **4**).

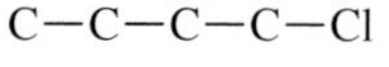

$$\mathrm{C{-}C{-}\overset{\delta\delta+}{C}{-}\overset{\delta+}{C}{-}\overset{\delta-}{Cl}}$$

4

Substituents in an aromatic ring that withdraw electrons in this way exert a **–I effect**. They include not only halogens and the hydroxyl and nitro groups, where an electronegative atom is attached to the ring, but also groups such as carbonyl and nitrile in which an electron-deficient carbon atom is bonded to the ring. Alkyl groups behave in the opposite manner, exerting a **+I effect** and releasing electron density to the ring.

The **mesomeric effect** is the analogous redistribution of electrons in π-bonds. However, this resonance effect is transmitted throughout the whole of a conjugated system and creates alternate polarity at the carbon atoms along the system. Substituents that withdraw electron density in this way (**–M groups**) include carbonyl (see **5**) and nitro groups, whilst electron-releasing (+M) functions include amino and hydroxy groups.

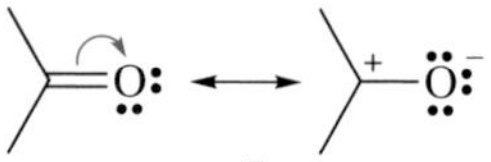

5

Note that some groups can withdraw electrons by one of the two effects but release electrons by the other, although one of the effects usually predominates.

2.3.1 Groups which Donate Electrons by the Mesomeric Effect

Groups (Z) in which the atom attached to the benzene ring possesses a lone pair of electrons can interact with the aromatic ring as shown in **6**, promoting *ortho* and *para* attack. The ring becomes more electron rich and so the reaction with electrophiles is facilitated. You can think of the lone pair of electrons as being formally located at the *ortho* and *para* positions.

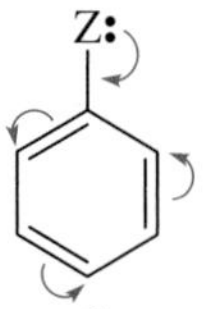

6

In order to assess the influence that substituents have on the reactivity of aromatic molecules, it is important to consider their effects not only on the benzene ring itself as above, but also on the carbocation intermediates resulting from electrophilic attack. These species are relatively unstable and any feature that affects their stability will influence their ease of formation and therefore the outcome of a reaction.

It is emphasized that none of these individual structures actually exist and that the structure of the intermediate carbocation is a hybrid of the various canonical forms.

We can illustrate the latter point by examining the attack by an electrophile E^+ on methoxybenzene (anisole) at the three possible sites of attack (Scheme 2.10).

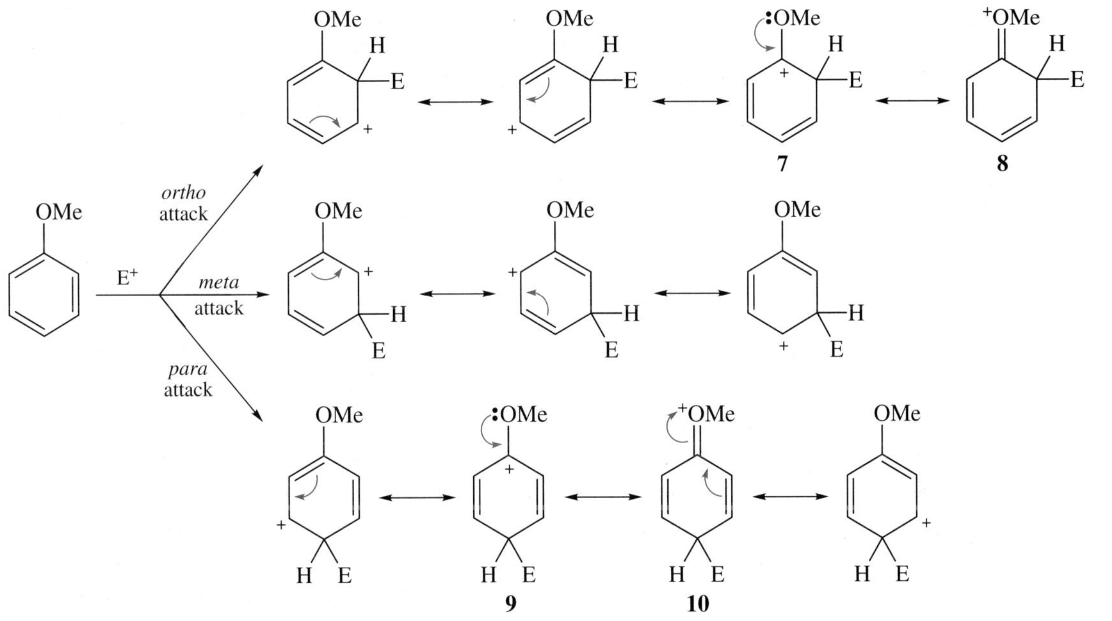

Scheme 2.10

Consider first attack at the *ortho* position. The structure **7** has the positive charge located on the carbon atom to which the methoxy group is bonded. Notice that this is a **tertiary carbocation**, a species that is recognized as being particularly stable (remember nucleophilic aliphatic substitution reactions). An additional canonical structure can be drawn involving donation of the lone pair of electrons on the oxygen atom to the electron-deficient C^+. This fourth canonical form confers extra stability on the intermediate and lowers the energy of the transition state leading to it. An oxonium species such as **8** is more stable than a carbocation, *e.g.* **7**, and hence can be considered to contribute more to the resonance hybrid.

Although **8** has been formally derived from **7**, and **10** from **9** (in Scheme 2.10), you should realize that involvement of the oxygen atom in the resonance stabilization could equally well have been shown from the other canonical forms for *ortho* and *para* attack.

A similar situation arises with species **9** associated with attack at the 4-position and this carbocation intermediate is therefore also additionally stabilized by **10**. However, no such structure can be drawn following *meta* attack and so the cation derived from this mode of attack is not additionally stabilized.

The consequences of the involvement of the methoxy group are to stabilize especially the carbocations arising from *ortho* and *para* attack and to lower the energy of activation for their formation, as illustrated in Figure 2.2. Notice that even attack at the *meta* position has a lower activation energy than does benzene.

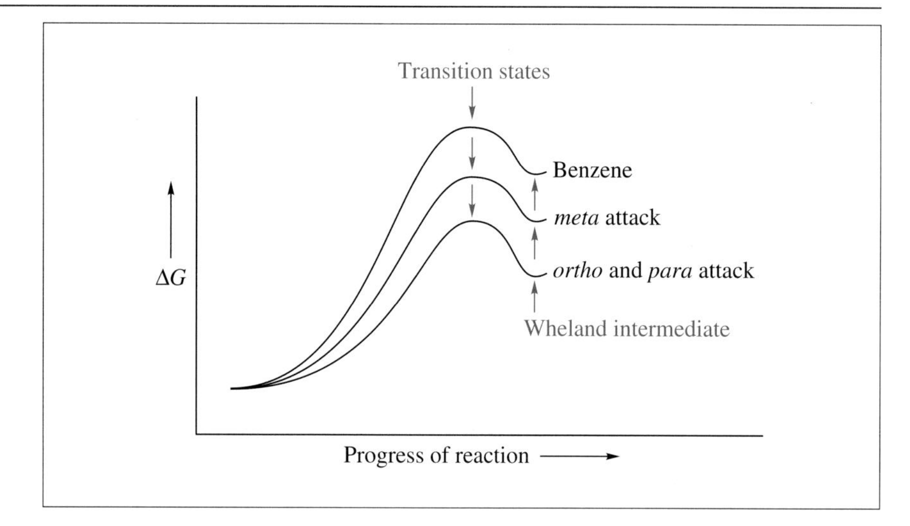

Figure 2.2 Energy profile for electrophilic attack on methoxybenzene at the *ortho*, *meta* and *para* positions compared with benzene

It should therefore be no surprise that the nitration of methoxybenzene is easier and faster than that of benzene and yields essentially only the 1,2- and 1,4-isomers (in almost equal amounts). Less than 1% of 3-nitroanisole is formed. Other electrophilic reactions follow this pattern.

Hydroxy and **amino** groups behave like a methoxy group. **Phenoxides**, in which the oxygen carries a full negative charge, are especially activated towards electrophilic attack.

The negative end of the dipole of chlorobenzene is on the chlorine atom, whereas in methoxybenzene the oxygen atom is the positive end of the dipole, supporting the view that overall a halogen is an electron-withdrawing substituent:

Halogen atoms also fall into this category. Possessing a lone pair of electrons, they are able to stabilize the intermediate cation arising from *ortho*/*para* attack. However, the halogenobenzenes behave differently from methoxybenzene and aniline in that the reaction with electrophiles is slower than for benzene. The nitration of chlorobenzene is about 30 times slower than that of benzene. Halogens are deactivating substituents and yet are *ortho*/*para* directors. As with methoxy and amino groups, the halogens withdraw electrons inductively, but donate them by the mesomeric effect. Only in the case of the halogens does the former effect dominate, with the consequence that the three intermediates from *ortho*, *meta* and *para* attack are all less stable than that arising from electrophilic attack of benzene. Nonetheless, *ortho*/*para* attack is still favoured because of the additional stabilization of the cations from the resonance forms **11** and **12**.

11 **12**

2.3.2 Groups which Withdraw Electrons by the Mesomeric Effect

Substituents which fall into this category include NO_2, CO_2R, COR, CN and SO_3R. All are characterized by the atom attached to the ring being linked to a more electronegative atom by a multiple bond and may be represented by X=Y, where Y is more electronegative than X (see Scheme

2.11). Electrons are therefore attracted towards Y, making X more electron deficient and therefore more strongly electron withdrawing. Formally, a positive charge is placed on the *ortho* and *para* positions.

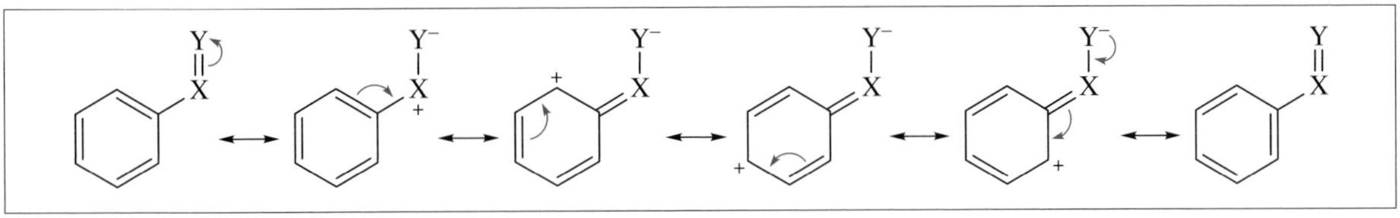

Scheme 2.11

Electrophilic attack on compounds which contain a substituent that withdraws electrons from the ring always leads to the 3-substituted compound, with very little of the 2- and 4-isomers being formed. The reaction is more difficult than for benzene, in keeping with the reduced electron density at the ring carbon atoms.

Again, it is important to examine the intermediates formed by attack of an electrophile, E^+, at the *ortho, meta* and *para* positions (Scheme 2.12). This time, nitrobenzene will be used as the substrate. It should be noticed that in the structures **13** and **14** associated with *ortho* and *para* attack, a positive charge is placed on the carbon to which the substituent is attached. The resulting situation is destabilizing because positive charges are located on adjacent atoms.

Scheme 2.12

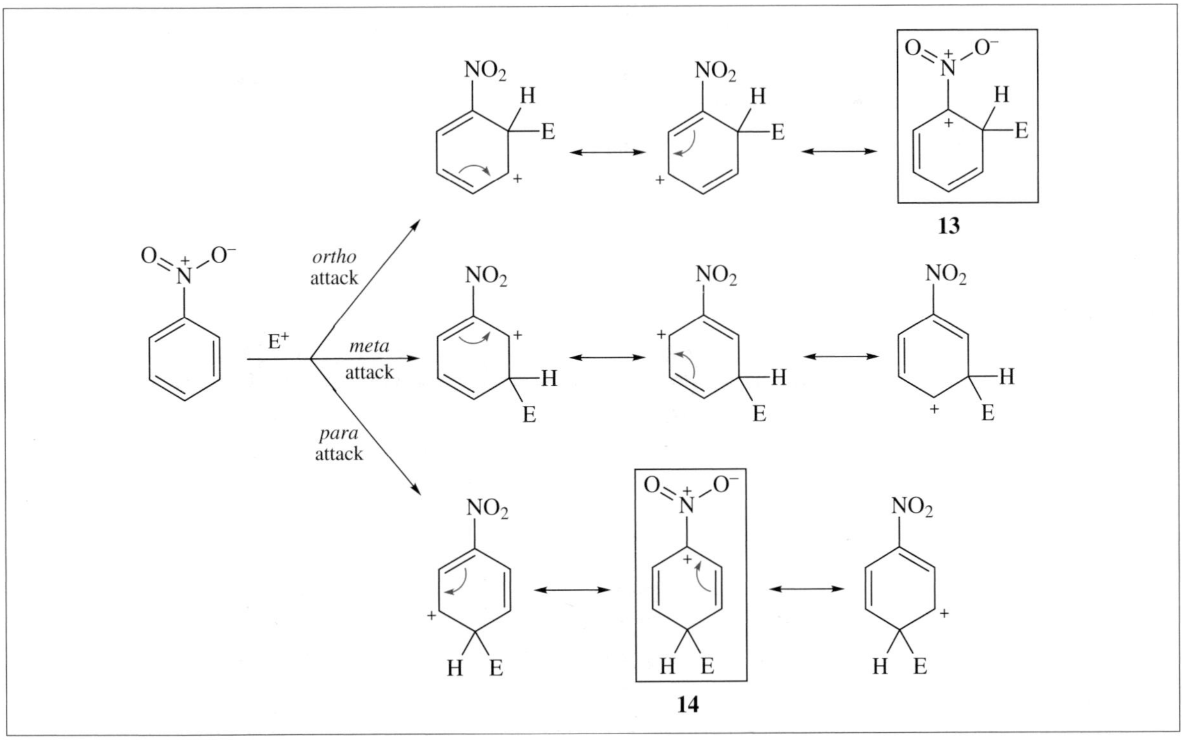

While attack at the 3-position is still much slower than for benzene, no canonical form places positive charges on adjacent atoms and so the intermediate is less destabilized than those arising from *ortho* and *para* attack. Hence *meta* attack is the preferred reaction, as illustrated in Figure 2.3. For example, nitration of nitrobenzene gives 88% of 1,3-dinitrobenzene and only 8% and 1% of the 1,2- and 1,4-isomers, respectively. The reaction occurs at a relative rate of 6×10^{-8} to that of benzene.

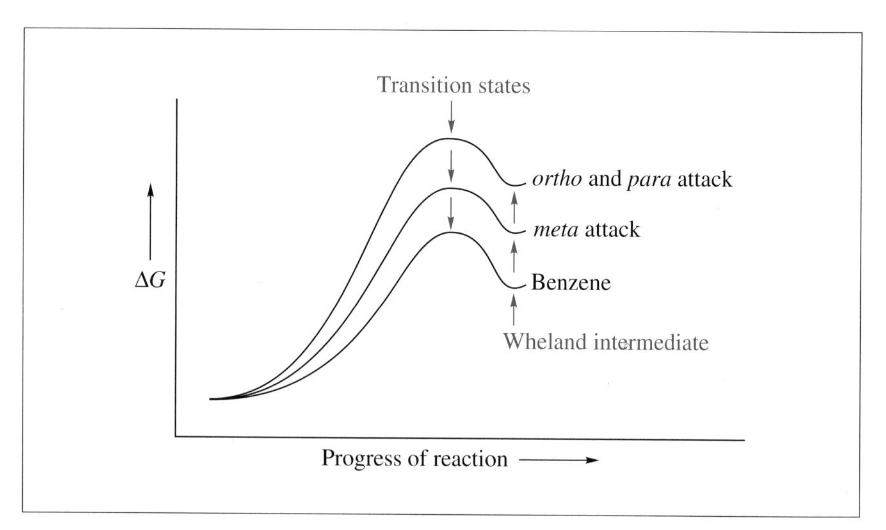

Figure 2.3 Energy profile for electrophilic attack on nitrobenzene at the *ortho*, *meta* and *para* positions compared with benzene

The efficiency of electron withdrawal by substituents increases in the order shown in Figure 2.4.

$Ar{-}C({=}O){-}OR < Ar{-}C({=}O){-}R < Ar{-}S({=}O)_2{-}OR < Ar{-}C{\equiv}N < Ar{-}\overset{+}{N}({=}O){-}O^-$

Figure 2.4

Worked Problem 2.2

Q Explain why a sulfonic acid group deactivates an aromatic ring to electrophilic substitution, which is directed to the 3-position.

A In benzenesulfonic acid, strongly electronegative oxygen atoms are bonded to the atom attached to the aromatic ring. Therefore electron density is withdrawn from the ring by both mesomeric and inductive effects. The canonical forms for the carbocation resulting from *para* attack by an electrophile are shown in Scheme 2.13. Application of the principles described in the chapter should enable you to draw the corresponding structures for the carbocations arising from *ortho* and *meta* attack. Examination of these canonical forms reveals that certain structures possess positive charges on adjacent atoms. These are high-energy canonical forms that contribute little to the resonance hybrid and effectively destabilize the carbocation. There are no such structures associated with *meta* attack, which is therefore the favoured site for reaction.

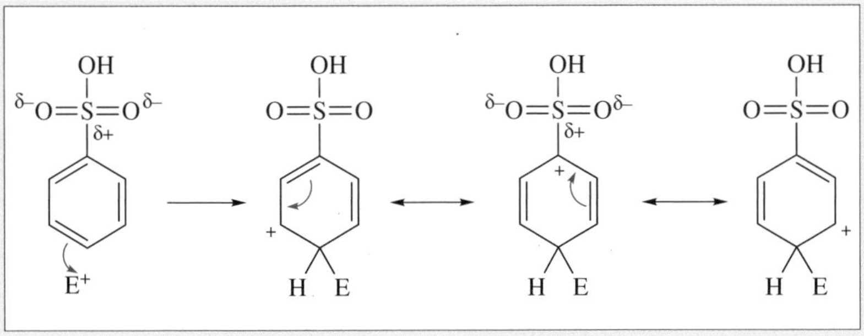

Scheme 2.13

2.3.3 Groups which Withdraw Electrons by the Inductive Effect

Groups such as **trifluoromethyl**, CF_3, and **trialkylammonium**, R_3N^+, are unable to interact with the π-system, but withdraw electrons as a result of the electronegativity of the fluorine atoms and the positively charged nitrogen, respectively. A study of the canonical forms for electrophilic attack at the three sites indicates a situation similar to that discussed above for mesomerically withdrawing groups (Scheme 2.14). The intermediates are overall destabilized by electron withdrawal, but structures **15** and **16** are particularly unfavourable because the positive charge is adjacent to the electron-deficient atom of the substituent. Thus, attack occurs preferentially at the 3-position, but is more difficult than electrophilic attack on benzene.

2.3.4 Groups which Donate Electrons by the Inductive Effect

It is well known that, in comparison to hydrogen, **alkyl groups** donate electrons. It is therefore to be expected that toluene and other alkylbenzenes will react with electrophiles rather more easily than benzene. This is certainly the case, toluene reacting with mixed acid at room temperature.

The canonical forms that contribute to the structure of the intermediate carbocation are shown in Scheme 2.15. Once again, one con-

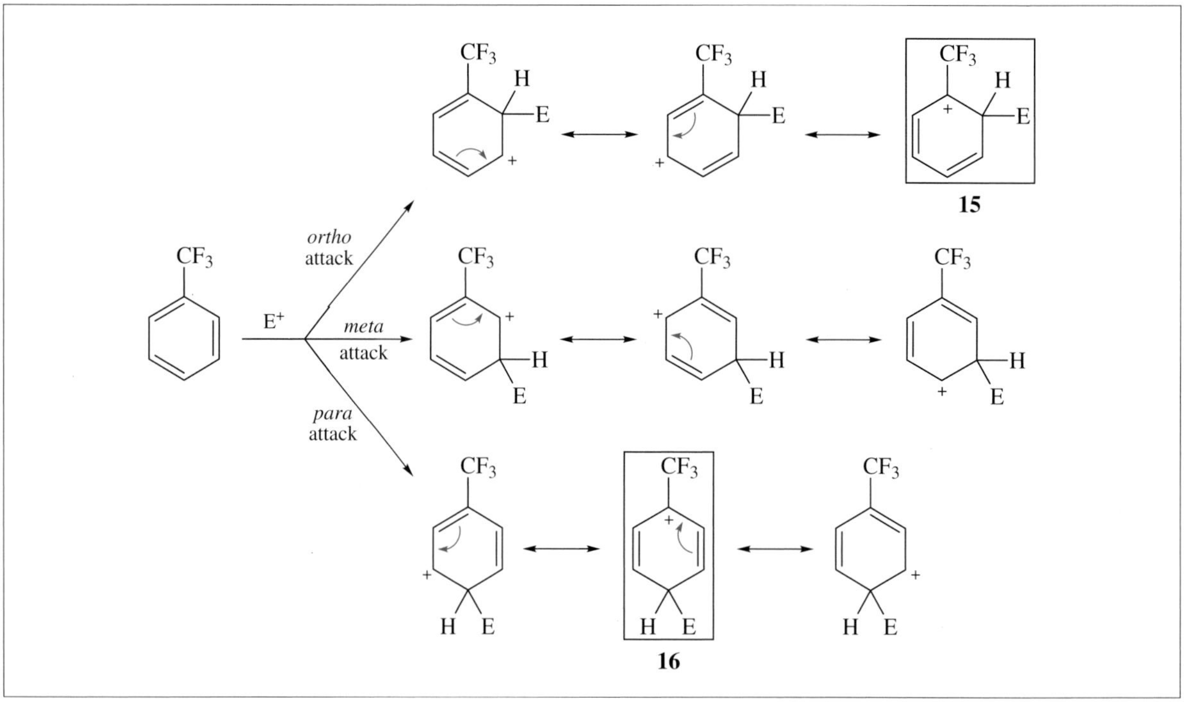
CF3
H
E
CF3
H
E
CF3
H
E
15
CF3
ortho
attack
E+
meta
attack
para
attack
CF3
H
E
CF3
H
E
CF3
H
E
CF3
H E
CF3
H E
16
CF3
H E

Scheme 2.14

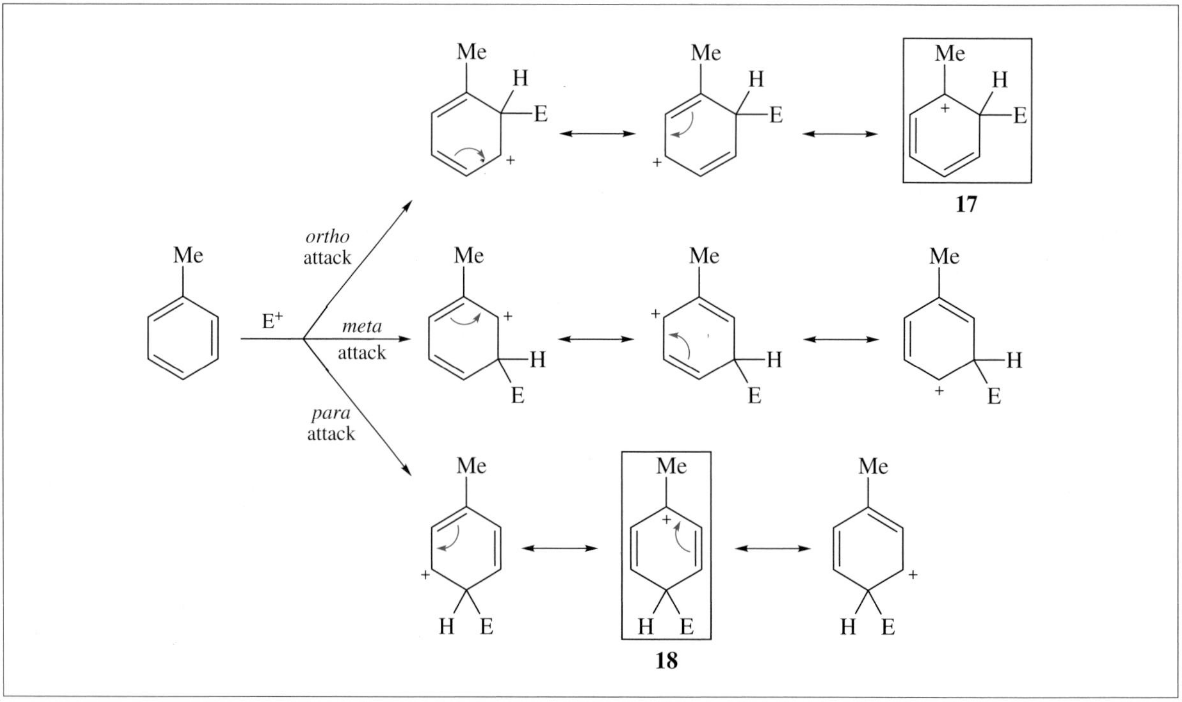
Me
H
E
Me
H
E
Me
H
E
17
Me
ortho
attack
E+
meta
attack
para
attack
Me
H
E
Me
H
E
Me
H
E
Me
H E
Me
H E
18
Me
H E

Scheme 2.15

tributing form derived from attack at the 2- and the 4-positions has the positive charge located on the carbon atom to which the substituent is attached. It is noted that these structures, **17** and **18**, are tertiary carbocations and that they are further stabilized by delocalization of the charge onto the methyl group, which therefore shares some of the electron deficiency. No such benefit results from attack at the 3-position, which is therefore not a favoured site for reaction. Nitration of toluene occurs about 25 times faster than that of benzene under similar conditions. It leads to a 2:1 mixture of 2- and 4-nitrotoluenes; only about 5% of the product is the 3-isomer (remember there are two *ortho* positions but only one *para* position).

The more efficient the alkyl group is at releasing electrons, the greater is the stabilization of the intermediate carbocation and the rate of electrophilic attack. Thus, *tert*-butylbenzene is nitrated faster than toluene.

This picture is somewhat generalized, since there are some exceptions. For instance, the chlorination of toluene proceeds faster than that of *tert*-butylbenzene. **Hyperconjugation** (Scheme 2.16) is at a maximum for a methyl group and has been offered as an explanation for these anomalies.

The stability of carbocations increases along the series:

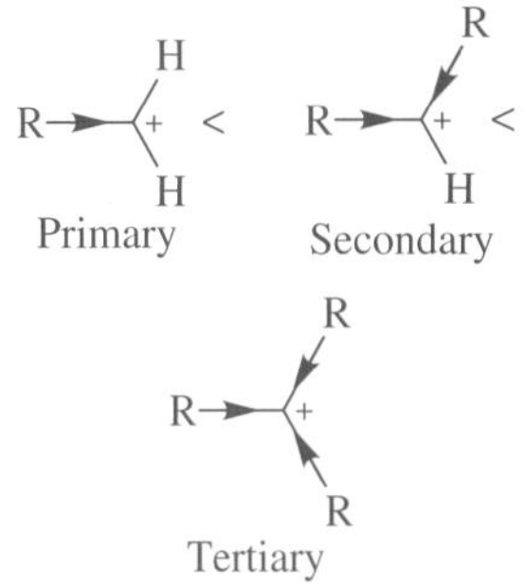

Primary < Secondary < Tertiary

The increasing number of electron-releasing methyl groups attached to the carbocation centre helps to spread the positive charge more effectively.

Hyperconjugation is a stabilizing interaction between a C–H bond and an adjacent σ-bond. For example, propene may be regarded as a resonance hybrid of two canonical forms:

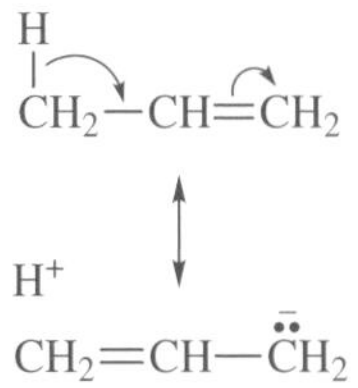

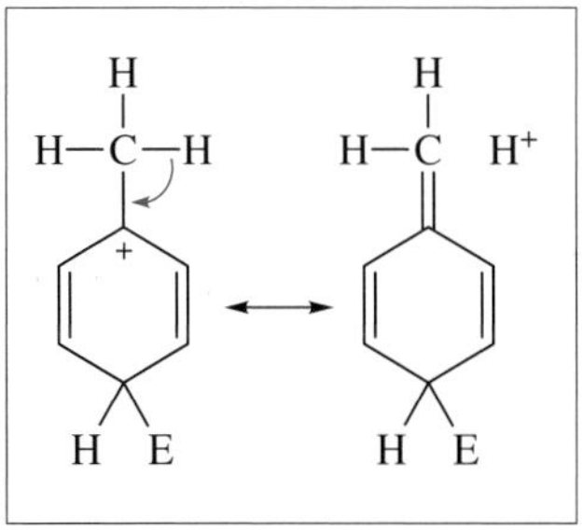

$H-CH_2-CH=CH_2 \leftrightarrow H^+ \; CH_2=CH-\ddot{C}H_2^-$

Scheme 2.16

Worked Problem 2.3

Q Identify the product when (3-chlorobutoxy)benzene is heated with $AlCl_3$ and account for its formation.

A This is an example of a Friedel–Crafts alkylation. The usual alkyl halide reagent is here part of the aromatic molecule that is attacked. The reaction is an example of an intramolecular process. The Lewis acid, $AlCl_3$, polarizes the halide side-chain, generating the attacking electrophile. It is not certain that a carbocation is formed; a complex between the substrate and $AlCl_3$ may be the attacking

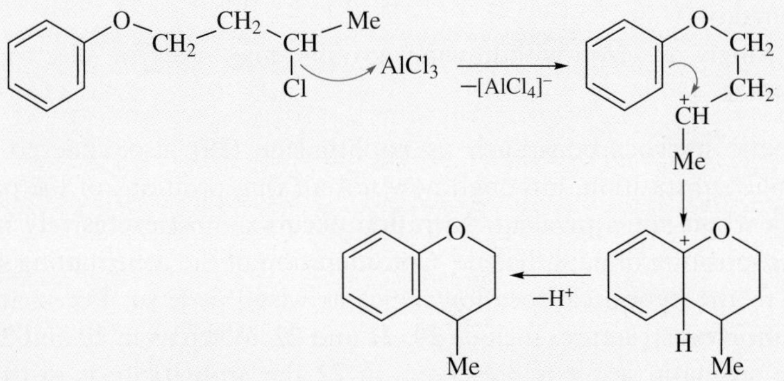

$-[AlCl_4]^-$; $-H^+$

species. The substituent is attached to the aromatic ring through an oxygen atom and the whole side-chain can be regarded as an alkoxy group. It is an electron-donating substituent and this helps electrophilic attack, which is directed to the *ortho* and *para* positions. However, the length of the side-chain is insufficient to reach the 4-position and so attack is restricted to the 2-position. The product is an oxygen heterocyclic compound, 4-methyl-3,4-dihydro-2*H*-1-benzopyran.

2.3.5 The Effects of Multiple Substitution

In general, the effects of two substituents on the orientation and rate of electrophilic substitution are additive. The best product selectivity occurs when the two substituents are working together, but unfortunately this is not always the case.

There are several guiding principles that help to decide the product in less obvious cases:

- Strongly activating groups dominate all other substituents
- Weakly activating groups next take control of orientation
- Deactivating groups exert the least control
- Steric effects often play a part in deciding the outcome of a reaction

When devising a synthesis of a particular compound (the target molecule), the effects of substituents have to be taken into account. It is essential to introduce substituents in the correct order so that their directing influence assists the synthesis rather than hinder it. Remember:

- *ortho/para* directors give mixtures of two isomers that can usually be separated
- *meta* directors give only the *meta* isomer
- *ortho/para* directing groups always overcome the influence of *meta* directors
- strongly electron-withdrawing groups may prevent electrophilic attack

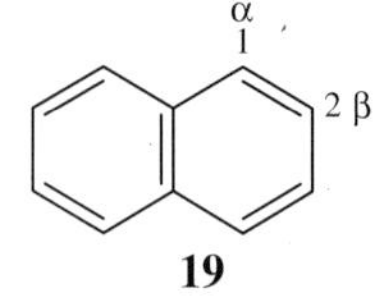

19

Aromatic hydrocarbons such as naphthalene (**19**) also undergo electrophilic substitution, although now not all ring positions of the parent hydrocarbon are equivalent. Nitration occurs almost exclusively in the 1- or α-position of naphthalene. Consideration of the contributing structures to the hybrid carbocation indicates why this is so. For α-attack, the canonical structures include **20**, **21** and **22**. Whereas in **20** and **21** the stable aromatic sextet is preserved, in **22** the aromaticity is disrupted. However, for attack at the 2- or β-position, only one structure, **23**, can be drawn in which the aromatic sextet is preserved. It is therefore expected that the carbocation produced during α-attack is more stable than that formed from β-attack and hence the rate of reaction at the α-position is significantly faster.

The effects of substituents on the regioselectivity of electrophilic aromatic substitution are summarized in Table 2.1.

A reaction which gives only one of several possible isomers is said to be regioselective.

Table 2.1 Reactivity and directing effects of substituent groups

ortho/para directors	*meta directors*
Strongly activating groups	Strongly deactivating groups
NR_2, NHR, NH_2, $NHCOCH_3$	NO_2, $^+NR_3$
O^-, OH, OR	SO_3H,
Weakly activating groups	CO_2H, CO_2R, COR
Alkyl, phenyl	CN, CF_3
Weakly deactivating groups	
F, Cl, Br, I	

2.4 The Hammett Equation

The relative ability of substituents in an aromatic ring to donate or withdraw electrons is indicated qualitatively by **Hammett substituent constants.** It was observed that a plot of the logarithms of the rate constants (k) for the alkaline hydrolysis of esters of benzoic acid against the pK_a values of the corresponding acids, $XC_6H_4CO_2H$, was linear, *i.e.*

$$\log k = \rho \times pK_a + C \tag{2.1}$$

where ρ (rho) and C are constants.

The line

$$\log k_0 = \rho \times pK_{a0} + C \quad (2.2)$$

describes the point for the unsubstituted compounds (X = H).

Subtraction of equation (2.2) from equation (2.1) gives:

$$\log k/k_0 = \rho \times (pK_{a0} - pK_a) \quad (2.3)$$

Log k and pK_a are related to free energies of activation and ionization, respectively, and hence a linear free energy relationship exists between the rates of ester hydrolysis and acid strengths.

Similar correlations between rate and equilibrium constants exist for various other side-chain reactions of benzene derivatives. The magnitude of ρ, which is called the **reaction constant**, is the slope of the line and varies with the reaction. The sign of ρ can be positive or negative according to whether the reaction rate is increased or decreased by the withdrawal of electrons.

The term ($pK_{a0} - pK_a$) is given the symbol σ (sigma) and is constant for given substituents. Equation 2.3 thus simplifies to:

$$\log k/k_0 = \rho\sigma \quad (2.4)$$

This is the **Hammett equation**.

The data for the ionization of benzoic acid and its derivatives in water at 25 °C are extensive and accurate and this was chosen as the standard reaction to which all other reactions would be compared. The value of ρ for the standard reaction is 1.00.

The **Hammett substituent constant**, σ, is a measure of the electron-donating or electron-withdrawing power of the particular substituent, with H being given a value of 0.00. Some typical values are listed in Table 2.2. These linear free energy correlations only apply to *meta* and *para* substituents in aromatic systems, since *ortho* substituents exert steric

Table 2.2 Hammett substituent constants for some substituents

Substituent	σ_{meta}	σ_{para}	*Substituent*	σ_{meta}	σ_{para}
O^-	−0.71	−1.00	F	+0.34	+0.06
NH_2	−0.16	−0.66	Cl	+0.37	+0.23
OH	+0.12	−0.37	$COCH_3$	+0.38	+0.50
CH_3	−0.07	−0.17	CF_3	+0.43	+0.54
OCH_3	+0.11	−0.27	CN	+0.56	+0.66
H	0.00	0.00	NO_2	+0.71	+0.78

effects which can alter the normal electronic behaviour. It can be seen that the more negative the value, the higher the electron-donating capacity of the group; substituents with a positive σ value are electron withdrawing.

The σ values reflect the interaction of the substituents with the reaction centre. The methoxy group can exert only its –I effect in the *meta* position; the stronger +M effect dominates in the *para* position. Consequently, σ_{meta} and σ_{para} have opposite signs for this group, indicating its electron-withdrawing and electron-donating ability, respectively.

2.5 Nucleophilic Aromatic Substitution

There are two distinct and major mechanisms by which a **nucleophile** can be introduced into the aromatic ring. In one, the nucleophile attacks at a ring carbon atom and this type is covered in detail below. The second method depends on an electron-rich species behaving as a base and attacking at hydrogen. This type of reaction is covered in detail in Chapter 9 and is only briefly considered here.

2.5.1 By an Addition–Elimination Mechanism (S_NAr)

Whereas electrophilic attack of benzene is both well known and important, the corresponding reaction with nucleophiles is very difficult and is not typical of aromatic compounds. However, if the aromatic ring is π-electron deficient because an electron-withdrawing group (EWG) is present, then nucleophilic attack can occur. The mechanism for the addition–elimination sequence for nucleophilic substitution is shown in Scheme 2.17.

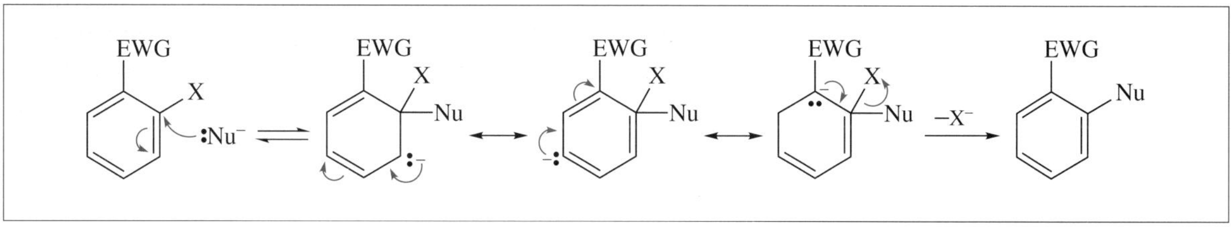

Scheme 2.17

The initial attack disrupts the π-cloud and the resulting intermediate species, a carbanion, is stabilized by resonance. There is a close similarity between this mechanism and that proposed earlier in this chapter for electrophilic attack on benzene, although in that reaction the intermediate was a carbocation. In both cases, this first step is usually the slower and therefore rate determining.

Evidence to support this mechanism for nucleophilic substitution includes isolation of several examples of the intermediate species and their structural determination by both NMR spectroscopy and X-ray crystallography. The intermediate species are called **Meisenheimer complexes**, many of which are strongly coloured.

The first such intermediate was isolated by Meisenheimer in 1902 from the reaction of 2-ethoxy-1,3,5-trinitrobenzene with sodium methoxide:

OEt
O_2N NO_2
NO_2
MeO^-Na^+
EtO OMe
O_2N NO_2
Na^+
NO_2

Compare the structure of this product with that of **3** (p. 17).

In the second step, aromaticity is restored through elimination of an ionic species and it is here that the two reaction types diverge:

- In electrophilic attack a proton, H^+, is lost through abstraction by a base.
- In nucleophilic substitution the leaving group is X^-.

The nature of X^- is of fundamental importance to the success of the reaction. Displacement of hydrogen is very difficult, because the hydride ion, H^-, is a very poor leaving group. Benzene itself does not react with nucleophiles. The important nucleophilic substitution reactions involve the displacement of a group other than hydride ion. The more effective leaving groups are the halogens, whose reactions are studied in Chapter 9, the diazonium group (Chapter 8) and the sulfonic acid group (Chapter 5). The diazonium group is the most effective because a very stable nitrogen molecule results from the elimination step. In these reactions, nucleophilic attack occurs at the carbon atom to which the substituent is attached. In the product, the nucleophile occupies the position of the original substituent. This process is called **aromatic nucleophilic *ipso* attack**. Electrophilic *ipso* substitution is discussed below.

Aryl halides only undergo substitution with extreme difficulty unless activated by electron-withdrawing groups, the role of which is to stabilize the intermediate species and so lower the energy of activation of the first step. In this respect, they serve the same purpose as donor groups in electrophilic substitution reactions. Nitro, nitrile and carbonyl are typical activating groups. Activation is best achieved when the group is *ortho* or *para* to the halogen, since both inductive and mesomeric withdrawal of electrons operate. The latter is of prime importance, providing additional resonance stabilization of the negative charge of the intermediate. This is illustrated for 1-chloro-2-nitrobenzene (Scheme 2.18) and is further discussed in Chapter 9. A 3-substituent is much less efficient at promoting nucleophilic attack since only the –I effect assists the process. Note that an electron-withdrawing substituent also reduces the electron density of the ring, thereby helping the initial attack by the nucleophile.

Electron-withdrawing substituents help nucleophilic substitution. Electron-donating groups help electrophilic reactions.

Pyridine is attacked directly by powerful nucleophiles. For example, 2-aminopyridine is synthesized by the reaction of pyridine with sodamide, $NaNH_2$, the **Tschitschibabin reaction**. The nitrogen heteroatom acts as a strongly electron-withdrawing group. Displacement of halogen from substituted pyridines is particularly easy, especially when *ortho* or *para* to the N heteroatom (see Heterocyclic Chemistry by M. Sainsbury in this series).

2.5.2 By an Elimination–Addition Mechanism

This topic is discussed in detail in Chapter 9 and only an outline is presented here. When simple aryl halides react with strong bases such as the amide ion, NH_2^-, a hydrogen atom adjacent to the C–halogen unit is abstracted by the base. The resulting carbanion acts as a nucleophile

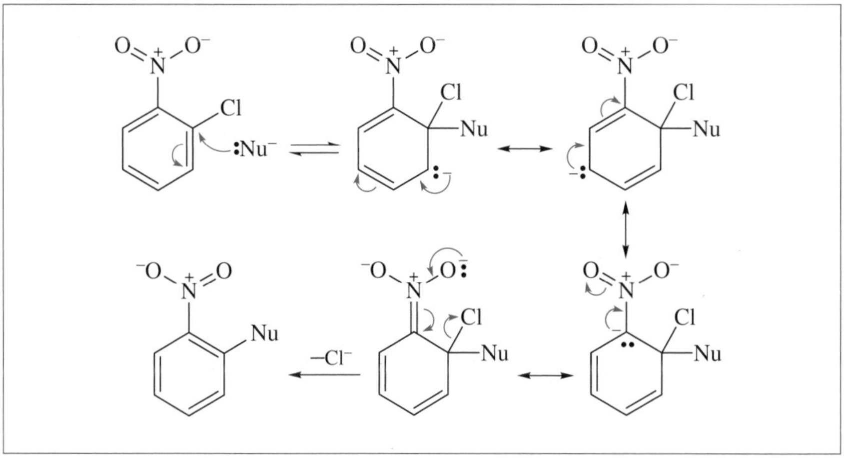

Scheme 2.18

and displaces the halide ion in an intramolecular process. The initial product is a highly reactive species called an aryne. It is rapidly attacked by NH_2^-, or its protonated derivative NH_3, now acting as a nucleophile. The final product, which results from protonation of a second carbanion, is the new substituted benzene derivative (Scheme 2.19).

An intramolecular reaction is one which occurs between two functional groups in the same molecule. In an intermolecular reaction, different molecules are involved.

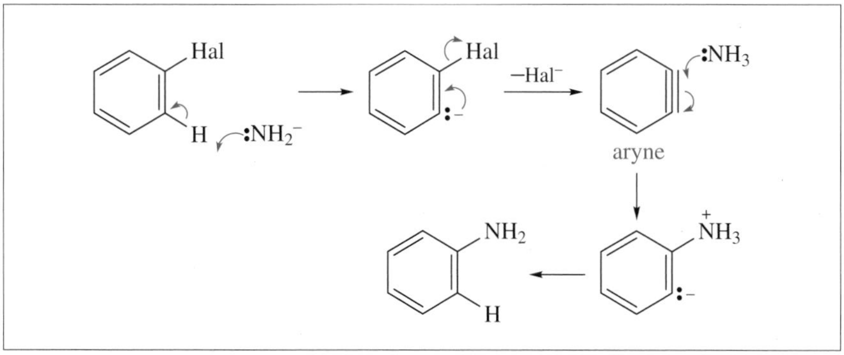

The common feature possessed by a base, B: or B:⁻, and a nucleophile, Nu: or Nu:⁻, is a lone pair of electrons. They differ in that the base donates the lone pair to an H atom, whereas donation to a C atom is an example of nucleophilic attack. Many species with a lone pair of electrons can act as both a base and a nucleophile.

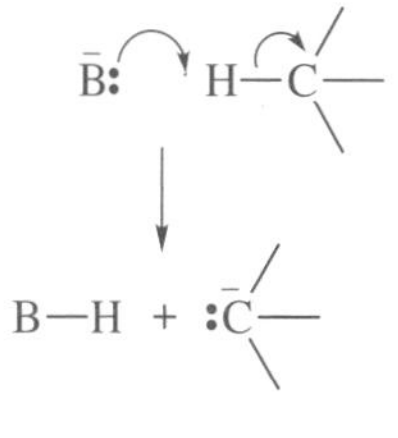

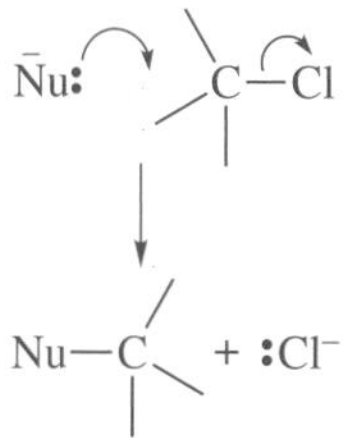

Scheme 2.19

2.6 *ipso* Substitution

Electrophilic attack can also occur at a position already occupied by a substituent, the *ipso* position. Such *ipso* substitutions are not common, but they are industrially useful. An example is *ipso* nitration by

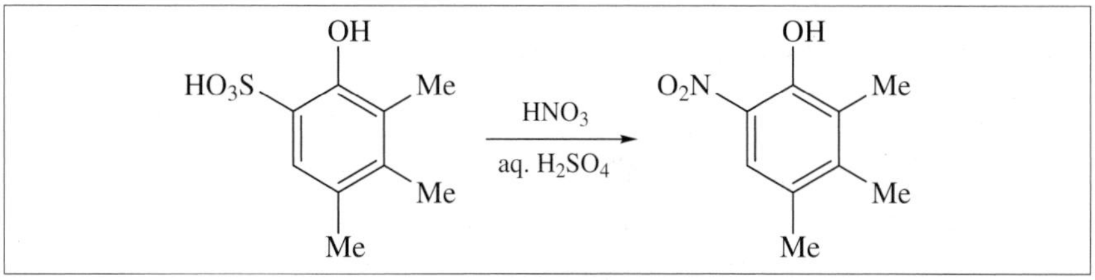

Scheme 2.20

displacement of a sulfonic acid group (Scheme 2.20). A proton can also displace the sulfonic acid group, with benzenesulfonic acid being converted into benzene. Nucleophilic *ipso* substitution reactions also occur (see Section 2.5.1).

Summary of Key Points

1. Electrophilic attack on benzene and related molecules proceeds by an addition–elimination mechanism. Initial attack generates a carbocationic intermediate from which loss of a proton restores the aromatic system. The carbocation intermediate is stabilized by resonance.

2. Electron-releasing substituents stabilize the positively charged intermediate and facilitate attack by an electrophile, which is directed to the *ortho* and *para* positions.

3. Electron-withdrawing substituents deactivate the ring to further electrophilic attack and direct the new substituents to the *meta* position.

4. Nucleophilic attack on benzene does not occur. Aromatic compounds possessing a good leaving group and containing strongly electron-withdrawing groups undergo nucleophilic substitution by an addition–elimination mechanism.

5. Arynes are generated from aryl halides by reaction with strong bases. The final outcome of the reaction is substitution of the halide.

Problems

2.1. Identify and account for the product(s) expected from each of the following reactions: (a) benzene + benzoyl chloride + $AlCl_3$; (b) acetanilide + nitric acid + sulfuric acid; (c) bromobenzene + sulfuric acid; (d) 1-methoxy-2-nitrobenzene + nitric acid + sulfuric acid; (e) (trifluoromethyl)benzene + Br_2 + Fe powder.

2.2. Suggest syntheses for each of the following compounds from the starting material indicated: (a) 3-bromoaniline from benzene; (b) 3-nitrobenzoic acid from toluene; (c) 4-nitrobenzoic acid from toluene; (d) 4-methylbenzophenone from benzoic acid; (e) 1-methoxy-2,4-dinitrobenzene from chlorobenzene.

2.3. Explain the following observations: (a) although Friedel–Crafts alkylation reactions often result in polysubstitution products, Friedel–Crafts acylation reactions only give the monosubstituted product; (b) bromination of ethyl 4-methylbenzoate gives only one product; (c) treatment of 4-chlorotoluene with $NaNH_2$ in liquid ammonia gives two products; (d) reaction of benzene with 1-chloro-2-methylpropane and $AlCl_3$ gives *tert*-butylbenzene.

3
Alkylbenzenes and Arylbenzenes

Aims

By the end of this chapter you should understand:

- The synthesis and reactions of arenes
- The Friedel–Crafts reaction

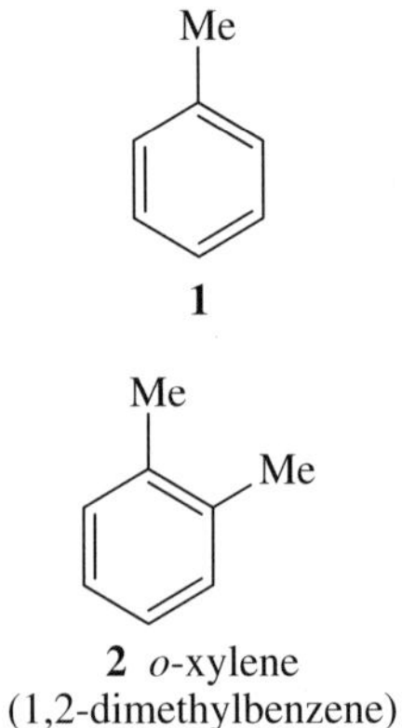

1

2 *o*-xylene (1,2-dimethylbenzene)

3 *m*-xylene (1,3-dimethylbenzene)

4 *p*-xylene (1,4-dimethylbenzene)

3.1 Introduction

Benzene and its simple alkyl derivatives are the building blocks of the aromatic chemical industry and are also important solvents for many reactions and processes. The simplest derivative, **toluene** (methylbenzene, **1**), is the source of a range of nitrotoluenes and is one of the most important industrial solvents. The three isomeric dimethylbenzenes, *o*-, *m*- and *p*-xylene (**2–4**) are often used as a mixture in industrial solvents.

3.2 Sources of Alkylbenzenes

Traditionally, the source of benzene and toluene has been coal. Coke is produced for use in the steel industry and a by-product of this process is coal tar which, when distilled, provides benzene, toluene, xylenes, phenol and cresols (methylphenols), and naphthalene, the most abundant single component.

However, the major source of these hydrocarbons is now petroleum. Although aromatic compounds do occur naturally in petroleum, they are mainly obtained by the process of catalytic reforming, in which aliphatic hydrocarbons are aromatized through dehydrogenation, cyclization and isomerization. The process, which is also known as **hydroforming**, is carried out under pressure at 480–550 °C in the presence of a catalyst, typically chromium(III) oxide or alumina. Benzene is thus produced from hexane, and toluene from heptane. Octane gives rise to the three isomers of xylene and to ethylbenzene. Since more toluene than benzene is produced in the process, a quantity of toluene is converted into benzene by **hydrodealkylation**. High temperature **cracking** (650–680 °C) of longer chain alkanes, a process that breaks them down into smaller alkanes, is also a source of aromatic compounds.

3.3 Introduction of Alkyl Groups

3.3.1 Friedel–Crafts Reaction

The most important means of introducing an alkyl group into an aromatic ring is the **Friedel–Crafts reaction**. In its simplest form, this is the reaction of an alkyl halide (halogenoalkane) with an aromatic compound, such as benzene, in the presence of a **Lewis acid**, commonly aluminium chloride (Scheme 3.1).

A Lewis acid accepts a pair of electrons.

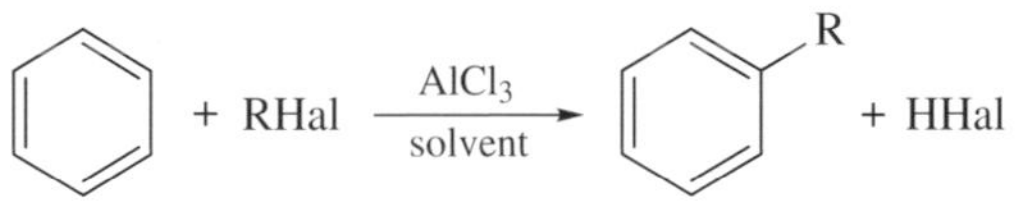

C6H6 + RHal —($AlCl_3$, solvent)→ C6H5R + HHal

Scheme 3.1

A wide range of reactants, catalysts, solvents and reaction conditions can be used, making the Friedel–Crafts reaction a very valuable and versatile process.

As well as alkyl halides, alcohols and alkenes are direct sources of alkyl groups. Acyl chlorides and anhydrides are additional sources, but these involve the subsequent reduction of a carbonyl group (C=O) to a methylene (CH_2) unit.

Reduction of carbonyl to alkyl can be achieved using zinc amalgam and hydrochloric acid (the Clemmensen reduction) and by hydrazine and a base (the Wolff–Kishner reduction):

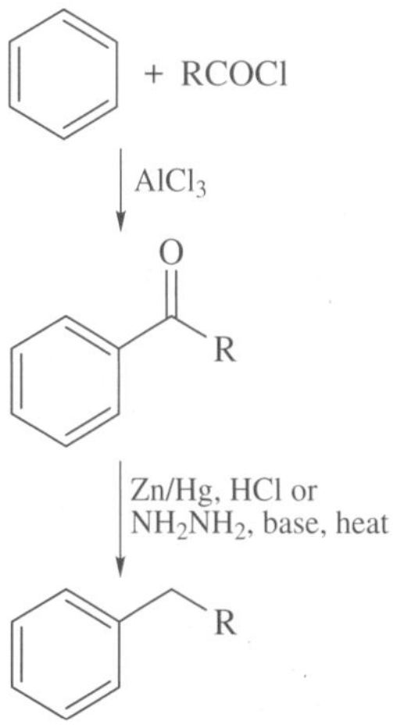

C6H6 + RCOCl —($AlCl_3$)→ C6H5COR —(Zn/Hg, HCl or NH_2NH_2, base, heat)→ C6H5CH2R

A variety of catalysts, including other Lewis acids such as $FeCl_3$ and BF_3, and the protic acids HF, phosphoric acid and sulfuric acid, has been used. In reactions using alcohols, the favoured catalyst is BF_3; HF is often used in reactions involving alkenes.

The reaction can be very fast, but can be moderated by the use of an inert solvent such as nitrobenzene or carbon disulfide. The temperature at which the reaction is carried out can vary from below room temperature to about 200 °C.

However, there are several drawbacks to this alkylation reaction. The use of longer alkyl chains than ethyl can be complicated by isomerization of the alkyl group arising from carbonium ion hydride shifts. It is therefore not uncommon for mixtures to be produced. In extreme cases, a completely different alkyl group from that of the starting material can be present in the product.

A specific example is the alkylation of benzene with 1-chloropropane

5 6

Isopropylbenzene, also known as cumene, is produced commercially by reaction of benzene with propene in the presence of sulfuric acid as catalyst.

in the presence of aluminium chloride. Propylbenzene (**5**) predominates when the reaction mixture is kept cold, but as the temperature is increased, isopropylbenzene (**6**) becomes the major product and at 80 °C accounts for approximately 70% of the mixture.

Reaction conditions can also influence the orientation of substitution. An example is the reaction of toluene with chloromethane in the presence of aluminium chloride. At room temperature, a mixture of 1,2- and 1,4-dimethylbenzenes results, but at 80 °C the product is mainly 1,3-dimethylbenzene. In fact, heating either of the 1,2- or 1,4-isomers in the presence of aluminium chloride/hydrochloric acid results in rearrangement to the more stable 1,3-dimethylbenzene.

A further drawback results from the electron-donating nature of alkyl groups, which assists attack on the benzene ring by electrophiles. The initial product, an alkylbenzene, is therefore more reactive than the starting material and a second and even further alkylation may occur, leading to mixed products.

Acylation by acid anhydrides also prevents isomerization, but this is wasteful of reagent, since half of it is converted into the carboxylic acid, and is therefore economically unacceptable. Nevertheless, cyclic anhydrides are viable reagents, as in the Haworth reaction (see Chapter 12).

Isomerization does not occur in the route that involves acylation and carbonyl reduction. This technique also prevents polysubstitution, since the acyl group is electron withdrawing and deactivates the ring to further electrophilic attack.

Friedel–Crafts alkylation fails when the substrate contains more powerful electron-withdrawing groups than halogen. Nitrobenzene is therefore a useful solvent for the reaction. Aromatic amines, although reactive towards electrophilic attack (see Chapter 8), do not undergo alkylation. The lone pair of electrons on the N atom of the amino group forms a coordinate bond to the $AlCl_3$, preventing its complexation to the alkyl halide. It should also be noted that the reaction does not work with aryl halides.

3.3.2 Mechanism of the Friedel–Crafts Reaction

The two mechanisms for Friedel–Crafts alkylation are not dissimilar to the two mechanisms for nucleophilic aliphatic substitution. In an S_N1 mechanism, a carbocation is generated from an alkyl halide before the nucleophile attacks, but in an S_N2 reaction the halide departs simultaneously with the nucleophile attacking the R group. In the Friedel–Crafts reaction, benzene behaves as the nucleophile.

In reactions involving alkyl halides, two mechanisms have been recognized which differ in the exact nature of the electrophile. One mechanism involves an **alkyl carbocation** generated by abstraction of the halogen from the alkyl halide by the $AlCl_3$. In the second process, a complex formed between the halide and the Lewis acid such as $[R-\overset{+}{Cl}-\overset{-}{Al}Cl_3]$ is the attacking electrophile. The difference between the two mechanisms is essentially whether an alkyl cation is actually formed (Scheme 3.2) and the effective mechanistic pathway may be somewhere between the two. In the first mechanism, the reaction involves attack of the cation on the benzene ring followed by abstraction of a proton by $[AlCl_4]^-$, which is formally converted into HCl and $AlCl_3$.

In the second mechanism, the alkyl group is transferred to the aromatic ring from the complex.

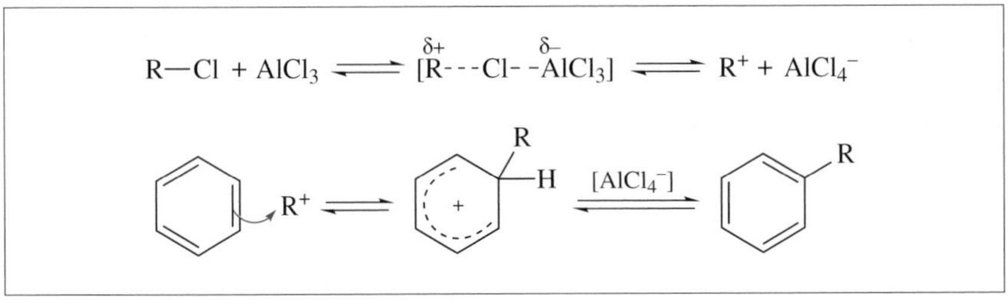

$$R{-}Cl + AlCl_3 \rightleftharpoons [\overset{\delta+}{R}{---}Cl{--}\overset{\delta-}{AlCl_3}] \rightleftharpoons R^+ + AlCl_4^-$$

Scheme 3.2

Alcohols and alkenes generate these carbocations in the presence of acids such as sulfuric acid (Scheme 3.3). Some alkyl cations rearrange to form the most stable ion, thus accounting for the isomerization noted earlier.

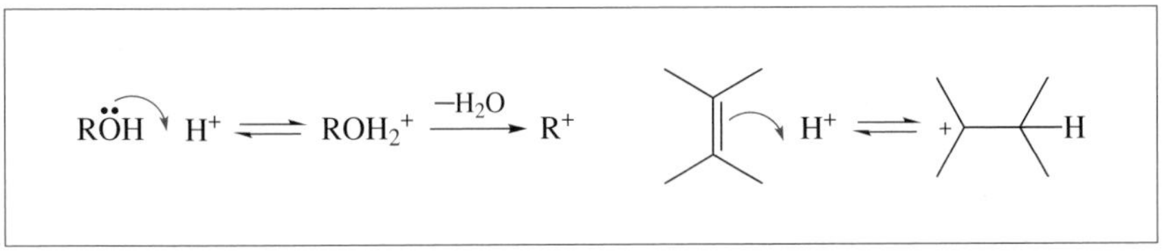

$$ROH + H^+ \rightleftharpoons ROH_2^+ \xrightarrow{-H_2O} R^+$$

Scheme 3.3

In the case of acyl chlorides, reaction with the Lewis acid generates an electrophilic **acylium ion**. These species show no tendency to rearrange (Scheme 3.4). Again, it is questionable whether a free cation is formed or if a complex between the acyl group and $AlCl_3$ is the attacking species.

Note the resonance stabilization of the acylium ion:

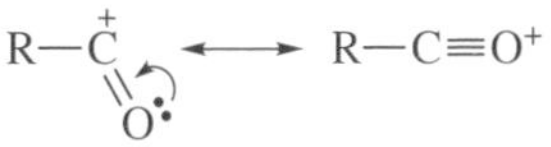

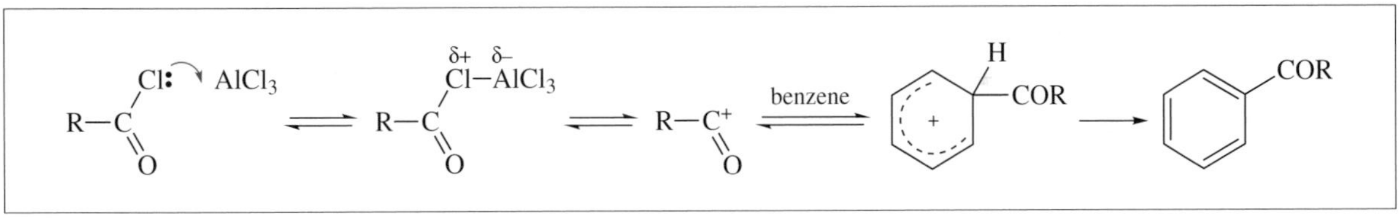

$$R{-}\overset{+}{C}{=}O \longleftrightarrow R{-}C{\equiv}O^+$$

Scheme 3.4

3.3.3 Wurtz–Fittig Reaction

Alkyl derivatives of benzene may be prepared by reacting an alkyl halide and an aryl halide with sodium in an inert solvent such as diethyl ether (Scheme 3.5). Although symmetrical by-products are also formed, it is possible to introduce long unbranched side-chains by this route without isomerization occurring.

It is also possible to synthesize alkyl derivatives of benzene using the Grignard reaction (see Chapter 10).

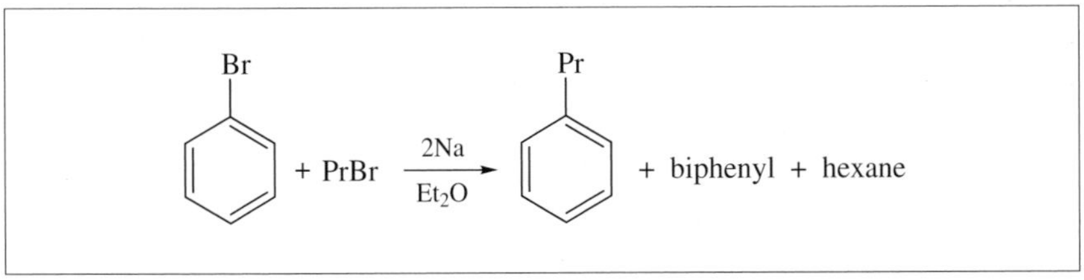

$$PhBr + PrBr \xrightarrow[Et_2O]{2Na} PhPr + \text{biphenyl} + \text{hexane}$$

Scheme 3.5

3.4 Reactions of Alkylbenzenes

3.4.1 Reactions of the Ring

An alkyl group activates the ring to electrophilic substitution mainly through an inductive effect and directs attack to the 2- and 4-positions. Examples of these reactions will appear throughout the book in the chapters on functionalized aromatic compounds.

3.4.2 Reactions of the Side-chain

Free Radical Halogenation

In the presence of light, but in the absence of a Lewis acid catalyst, halogenation of toluene occurs in the methyl group by a free-radical mechanism. The reaction proceeds stepwise, leading eventually to (trichloromethyl)benzene (benzotrichloride, $PhCCl_3$). With ethylbenzene, a similar reaction results; chlorination occurs initially at the α-position.

When an alkyl side-chain is attached to a benzene ring, the different carbon atoms are identified by Greek letters, with the atom attached to the ring being designated α. In a long-chain compound, the terminal carbon is often designated as ω:

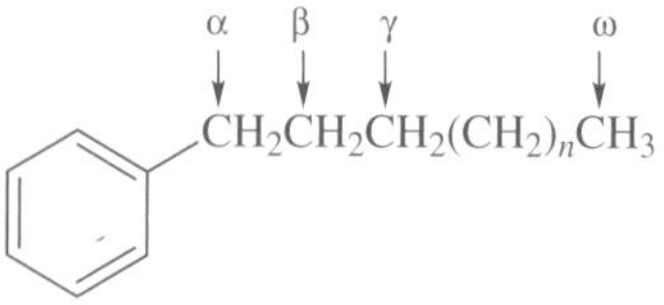

α β γ ω

$CH_2CH_2CH_2(CH_2)_nCH_3$

Cl

8

Worked Problem 3.1

Q Explain why ethylbenzene is attacked preferentially at the α-position by chlorine in the presence of light. Indicate the mechanisms of ring halogenation and side-chain halogenation.

A Ring halogenation and side-chain halogenation proceed by different mechanisms. It has been seen in Chapter 2, and is discussed in more detail in Chapter 9, that attack by halogen atoms in the presence of a Lewis acid catalyst proceeds by electrophilic substitution (see p. 18). However, in the presence of light, a free-radical mechanism operates and side-chain halogenation occurs, as shown here for toluene:

- Initiation $Cl_2 + h\nu \rightarrow 2Cl\cdot$
- Propagation $PhCH_3 + Cl\cdot \rightarrow PhCH_2\cdot + HCl$
- Propagation $PhCH_2\cdot + Cl_2 \rightarrow PhCH_2Cl + Cl\cdot$

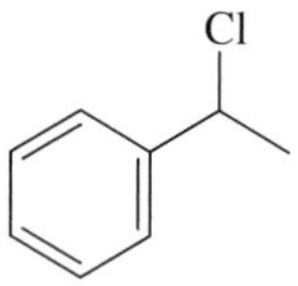

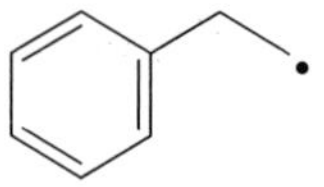

There are two different sites in ethylbenzene from which a hydrogen atom can be abstracted by a chlorine radical, leading to two possible radicals. The first radical **7** (Scheme 3.6) is stabilized by resonance interaction with the benzene ring. This is not possible with the second radical **8** and hence this mode of attack is not favoured. Therefore α-substitution leads to the more stable radical and so predominates.

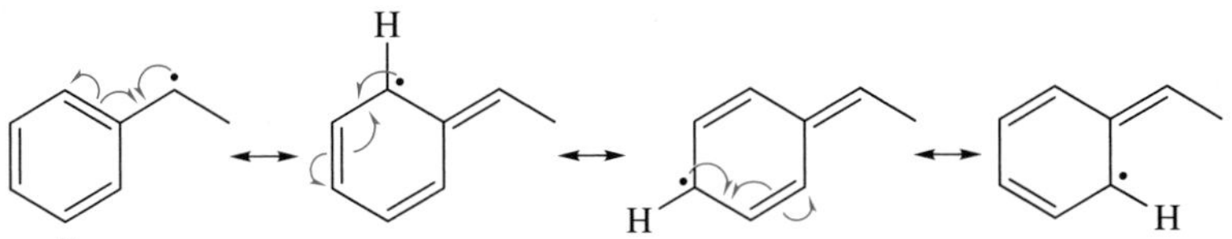

The movement of a single electron is shown by a fish-hook arrow:

Scheme 3.6

Side-chain Oxidation

Oxidation of aromatic systems containing alkyl side-chains results in the formation of a carboxylic acid, irrespective of the length of the side-chain. The usual oxidizing agents are potassium permanganate [potassium manganate(VII)] or chromic acid [chromium(VI) acid]. For example, 1,4-dimethylbenzene is oxidized to benzene-1,4-dicarboxylic acid (terephthalic acid, **9**), an important building block for polyesters. The oxidation of isopropylbenzene (cumene) to phenol is an important industrial process and is discussed in Chapter 4.

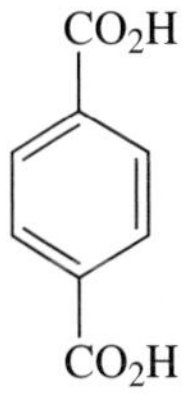

Side-chain Dehydrogenation

Styrene (phenylethene, **10**) is an important industrial chemical that is prepared by dehydrogenation of ethylbenzene at 600 °C over zinc oxide or chromium(III) oxide on alumina (Scheme 3.7). Ethylbenzene can be produced from benzene and ethene by a Friedel–Crafts reaction.

Styrene is used in the production of industrial plastics such as polystyrene. The side-chain undergoes the typical reactions of an unsaturated system.

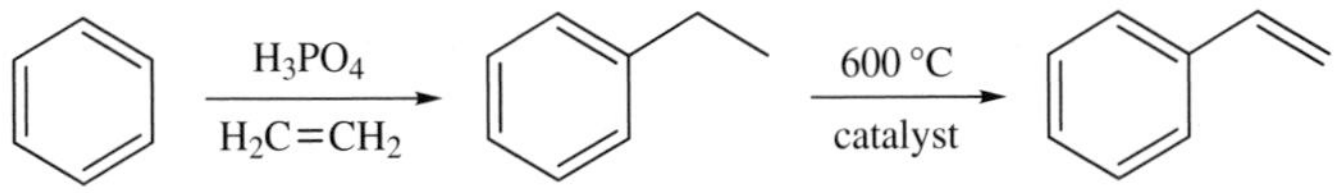

Scheme 3.7

Worked Problem 3.2

Q Dehydrohalogenation of (2-chloropropyl)benzene yields only 1-propenylbenzene. Explain.

A The reaction could produce two possible products through loss of either an α-proton or a γ-proton together with loss of the Cl^-

ion (Scheme 3.8). However, a compound which can undergo an elimination reaction in two directions usually gives the more highly substituted alkene (Saytzeff's rule), since this product is the more stable. Here, there is the additional factor that the double bond in prop-1-enylbenzene is conjugated with the aromatic ring. In fact, heating allylbenzene (prop-2-enylbenzene) results in its conversion to the more stable prop-1-enylbenzene.

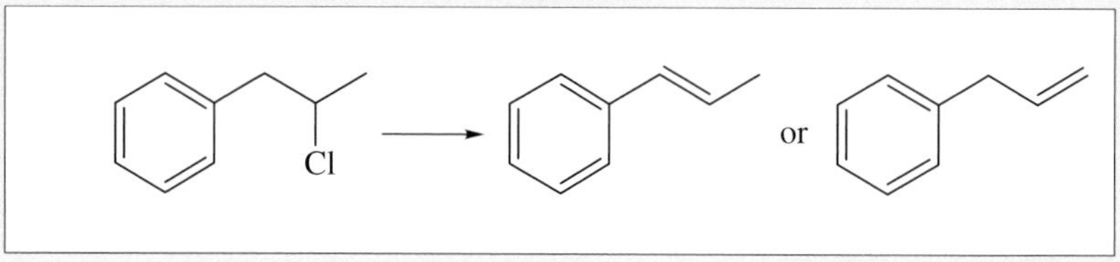

Cl — or

Scheme 3.8

3.5 Aryl Derivatives of Benzene

There are two classes of compounds that can be considered to fall into this category. The simplest such derivative is **biphenyl (11)**, in which two benzene rings are connected *via* a carbon–carbon single bond. This compound can be prepared by **Fittig's reaction** from bromobenzene or by the **Gomberg reaction** from benzenediazonium sulfate in the presence of ethanol and copper (Scheme 3.9), although yields are poor. Biphenyls may also be prepared using organometallic coupling (see Chapter 10).

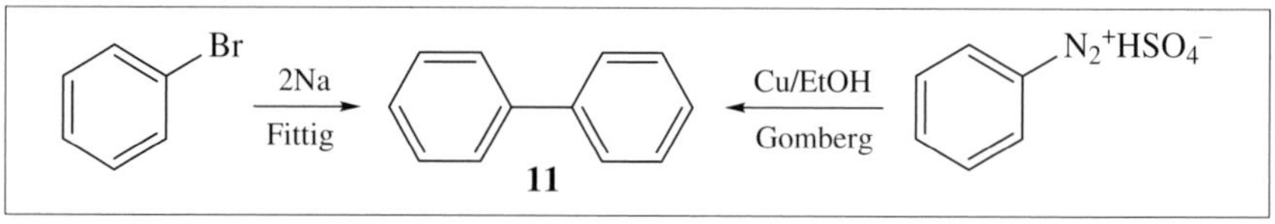

Br 2Na Fittig 11 Cu/EtOH Gomberg $N_2^+HSO_4^-$

Scheme 3.9

Biphenyl undergoes typical electrophilic substitution reactions. The phenyl group is *ortho*/*para* directing. For example, the major product of mononitration is 4-nitrobiphenyl. Introduction of a second nitro group in the molecule occurs in the unsubstituted ring, also, mainly, in the 4′-position. This might be unexpected since a nitrophenyl group is electron withdrawing, and therefore *meta* directing. However, irrespective of the electronic properties of the mono substituent, electrophilic substitution of a second substituent generally occurs in the 4′-position of the unsubstituted ring. The positive charge associated with the carbocation intermediate from *para* substitution can be delocalized into the second phenyl ring and so is efficiently stabilized. This is not the case with the Wheland intermediate from *meta* attack, which is therefore not the preferred site of substitution. You should draw these two possible intermediate cations and their resonance structures to confirm this.

The other type of compound considered here is where two benzene rings are bonded *via* a methylene bridge. The simplest such compound is diphenylmethane (**12**), which can be synthesized in a number of ways using Friedel–Crafts methodology. The reaction of an excess of benzene with dichloromethane in the presence of aluminium chloride results in the displacement of both halogen atoms. It can also be prepared from benzene and (chloromethyl)benzene, and from benzoyl chloride and benzene followed by reduction of the carbonyl group of the resulting benzophenone (Scheme 3.10).

Diarylethanes, $ArCH_2CH_2Ar$, can be prepared from 1,2-dichloroethane and arenes. Triphenylmethane, Ph_3CH, can be obtained by reacting an excess of benzene with trichloromethane under Friedel–Crafts conditions.

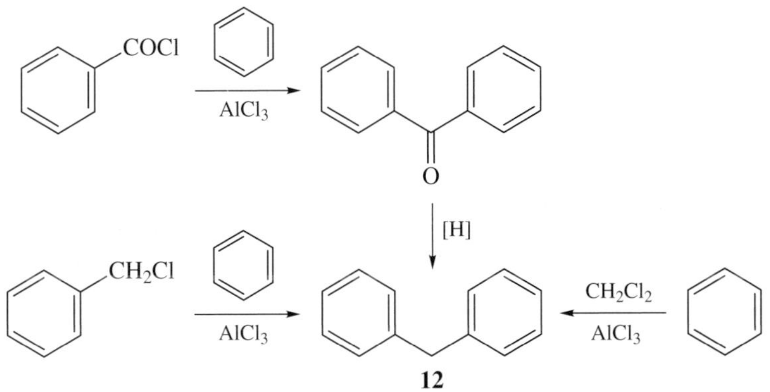

COCl

$AlCl_3$

O

[H]

CH_2Cl

$AlCl_3$

CH_2Cl_2

$AlCl_3$

12

Scheme 3.10

Reactions of diphenylmethane are similar to those of biphenyl, since the benzyl group, $C_6H_5CH_2$, is also *ortho*/*para* directing, although bromination results in reaction at the methylene group.

Summary of Key Points

1. Alkylbenzenes are prepared by the Friedel–Crafts reaction in which benzene is reacted with an alkyl halide in the presence of a Lewis acid such as $AlCl_3$. Diarylmethanes are formed in this manner from dichloromethane and triarylmethanes from trichloromethane.

2. Alkylbenzenes are attacked at the 2- and 4-positions by electrophiles. The reaction is easier than for benzene.

3. Biphenyls can be obtained from the reaction of bromobenzene with sodium and from diazonium salts.

4. In the absence of a Lewis acid but in the presence of UV light, halogenation of alkylbenzenes occurs in the side chain by a free-radical mechanism.

5. Oxidation of alkylbenzenes leads to aromatic carboxylic acids.

Problems

3.1. Identify the product from each of the following reactions: (a) chlorobenzene + propan-2-ol + conc. H_2SO_4; (b) bromobenzene + AcCl + $AlCl_3$; (c) benzene + $PhCH_2COCl$ + $AlCl_3$; (d) methoxybenzene + Ac_2O + BF_3 etherate; (e) benzene + butanedioic anhydride + $AlCl_3$.

3.2. Suggest syntheses of the following compounds from the starting material given: (a) 4-chloro-3-nitrobenzoic acid from toluene; (b) 2-chloro-4-nitrobenzoic acid from toluene; (c) 4-benzylbenzoic acid from toluene; (d) 1-ethyl-3-nitrobenzene from benzene; (e) isobutylbenzene from benzene.

4

Phenols

Aims

By the end of this chapter you should understand:

- How the hydroxyl group can be introduced into an aromatic ring
- The acidity of phenols
- The influence of the hydroxyl group on aromatic reactions

4.1 Introduction

Phenol (hydroxybenzene, **1**) has a hydroxyl group attached directly to the benzene ring. Phenol is a stable enol and, although there are some obvious similarities, the hydroxyl group exhibits sufficiently different properties from an alcoholic hydroxyl group to merit a separate classification.

OH OH Me

1 **2**

Phenol, previously known as carbolic acid, is both corrosive to human tissue and poisonous and care is needed in its handling. It is an important industrial compound with many uses, a major one being in the production of phenol–formaldehyde polymers such as Bakelite.

4.2 Industrial Synthetic Methods

Although coal tar is still an industrial source of phenol and the three cresols (methylphenols), *e.g.* *m*-cresol (**2**), and the dimethyl derivatives (xylenols), synthetically manufactured material predominates.

Most phenol nowadays is obtained from isopropylbenzene (cumene), which is oxidized by air in the **cumene process** (Scheme 4.1). Acetone (propanone) is a valuable by-product of the process and this route is a major source of this important solvent. The formation of cumene hydroperoxide proceeds by a free radical chain reaction initiated by the ready generation of the tertiary benzylic cumyl radical, which is a further illustration of the ease of attack at the benzylic position, especially by radicals (see Chapter 3).

Cumene is obtained from benzene by Friedel–Crafts alkylation with propene.

Hemiacetals or hemiketals are usually formed by the acid-catalysed addition of an alcohol to the carbonyl group of an aldehyde or ketone, e.g.

$$\mathrm{MeC(=O)H + MeOH \overset{H^+}{\rightleftharpoons} MeC(OH)(OMe)H}$$

The mechanism is considered to proceed as shown in Scheme 4.1. Protonation of the cumene hydroperoxide results in loss of water, generating an electron-deficient oxygen atom. A 1,2-shift of the phenyl group occurs, probably simultaneously. Finally, the protonated hemiketal **3** is hydrolysed under the acidic conditions to produce phenol and acetone.

Chlorobenzene, commercially produced by the **Raschig process** (see p. 108), is resistant to nucleophilic substitution under normal conditions, but in the **Dow process**, treatment with sodium hydroxide at 300 °C under high pressure is effective. Phenol may also be prepared from chlorobenzene by reaction with steam at 450 °C over a catalyst.

$$\mathrm{PhCHMe_2 \xrightarrow{R^\bullet} Ph\dot{C}Me_2 \xrightarrow{O_2} PhCMe_2{-}O{-}O^\bullet \xrightarrow{PhCHMe_2} PhCMe_2{-}O{-}OH \overset{H^+}{\rightleftharpoons} PhCMe_2{-}O{-}\overset{+}{O}H_2}$$

$$\mathrm{\overset{-H_2O}{\rightleftharpoons} Me_2\overset{+}{C}{-}OPh \overset{H_2O}{\rightleftharpoons} Me_2C(\overset{+}{O}H_2)(OPh)\ (\mathbf{3}) \overset{-H^+}{\rightleftharpoons} PhO^- + Me_2C{=}\overset{+}{O}H \longrightarrow PhOH + Me_2C{=}O}$$

Scheme 4.1

4.3 Laboratory Syntheses

The hydroxyl group cannot be directly substituted into the aromatic ring, but is introduced through conversion of other substituents.

4.3.1 From Arenesulfonic Acids

The fusion of alkali metal sulfonates with alkali in the presence of some water is used both in the laboratory and in industry (Scheme 4.2).

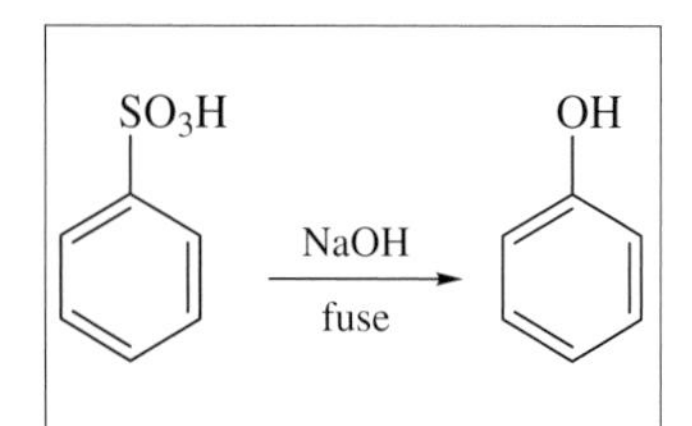

Scheme 4.2

4.3.2 From Aryl Halides

There are two useful ways by which halogen can be displaced by a hydroxyl group. As discussed in Chapters 2 and 9, only aryl halides that are activated by electron-withdrawing groups are susceptible to nucleophilic substitution, when even water and aqueous sodium hydroxide can be effective reagents. A hydroxyl group can also be introduced by conversion of the aryl halide into a Grignard reagent or an aryllithium compound and subsequent reaction with oxygen or through the intermediacy of the boronic acid (see Chapter 10).

4.3.3 From Amino Compounds

The amine is converted into a diazonium salt which is then warmed with water (see Chapter 8).

4.3.4 Miscellaneous Methods

There are a number of less frequently used methods for the preparation of phenols that are worthy of mention. The rearrangement of 2-hydroxybenzaldehydes brought about by reaction with alkaline hydrogen peroxide and leading to dihydroxybenzenes (the **Dakin reaction**) is discussed in Section 4.8. The acid-catalysed rearrangement of phenylhydroxylamines, known as the **Bamberger rearrangement**, is useful for the synthesis of 4-aminophenols (Scheme 4.3).

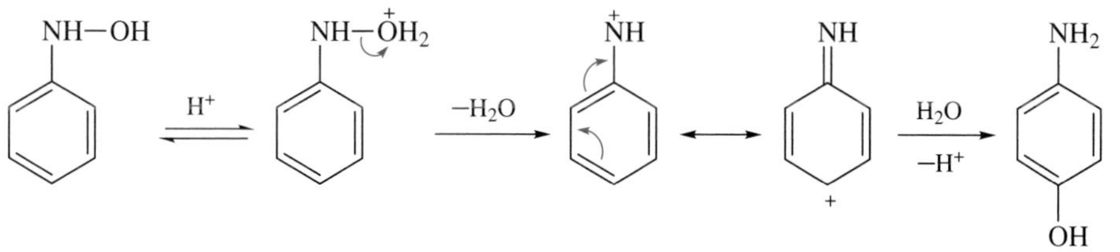

NH—OH; H^+; $NH-\overset{+}{O}H_2$; $-H_2O$; $\overset{+}{N}H$; NH; +; H_2O; $-H^+$; NH_2; OH

Scheme 4.3

Worked Problem 4.1

Q Suggest mechanisms for the displacement of the sulfonic acid group by hydroxide ion and the acid cleavage of ethers to produce phenols, and comment on the differences.

A In the displacement of the sulfonic acid group, the attacking species is negatively charged and therefore attacks the most electron-deficient centre, which is the carbon atom closest to the sulfonic acid group and bonded to the sulfur atom. Thus the mechanism for sulfonic acid displacement is as shown in Scheme 4.4.

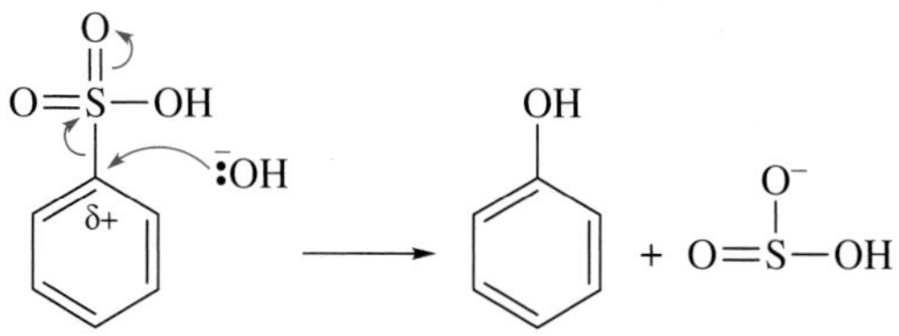

O; O=S—OH; $\bar{:}$OH; δ+; OH; + O=S—OH; O^-

Scheme 4.4

In the case of the acid cleavage of ethers, which can be achieved with either concentrated hydriodic acid or hydrobromic acid, this proceeds by protonation of the oxygen atom, followed by nucleophilic displacement of phenol as shown in Scheme 4.5.

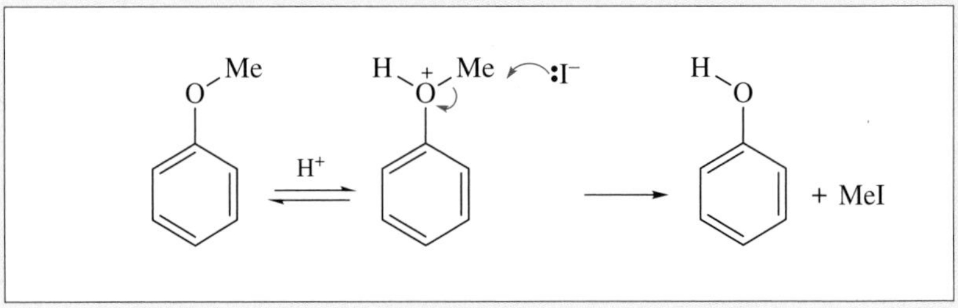

Scheme 4.5

4.4 The Acidity of Phenols

Phenols are converted into salts with strong alkalis such as sodium hydroxide, but not with sodium hydrogen carbonate solution. They are therefore stronger acids than alcohols but weaker than carboxylic acids. The pK_a of phenol is 9.95 compared with 4.20 for benzoic acid and about 17 for cyclohexanol.

The acidity of phenols arises from the greater resonance stabilization of the phenoxide anion compared with phenol itself (Scheme 4.6). There is no energy-demanding separation of charge in the resonance structures

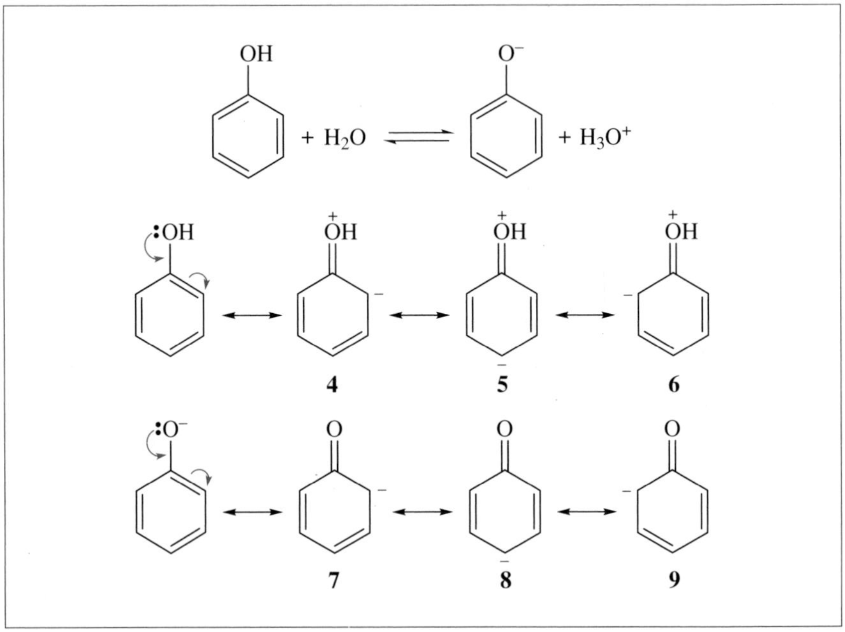

The strength of an acid HX is derived from the equilibrium: $HX + H_2O \rightleftharpoons H_3O^+ + X^-$

The equilibrium constant is given by:

$$K_a = \frac{[H_3O^+][X^-]}{[HX]}$$

K_a is converted into pK_a, where p$K_a = -\log_{10}K_a$. This means that a small value of K_a corresponds to a comparatively large value of pK_a, and *vice versa*. For example, $K_a = 10^{-14}$ mol dm^{-3} means a pK_a of 14. Hence from their pK_a values given in the text, phenol is 10^7 times more acidic than cyclohexanol.

The anion derived from cyclohexanol cannot be resonance stabilized and cyclohexanol is therefore a much weaker acid than phenol. Most alcohols have a pK_a value of about 17.

Scheme 4.6

for the anion (**7–9**) as there is for phenol (**4–6**). Thus, the equilibrium between phenol and its anion is displaced towards the latter species, with a corresponding increase in acidity.

The influence of ring substituents on the acidity of the phenolic group is dependent on the electronic properties of the substituent and its position in the ring relative to the hydroxyl group. Consider first the three mononitrophenols. In all cases, the electron-withdrawing inductive effect (–I) of the nitro group will cause an increase in acidity. However, the nitro group can also interact mesomerically with the hydroxyl group when it is in the 2- and 4-positions. The increased stabilization arising from the –M effect, illustrated by the contributing structure **10**, has a marked effect on the acidity. Thus, both 2- and 4-nitrophenols (pK_a 7.23 and 7.15, respectively) are approximately 1000 times stronger acids than phenol. They are more than 10 times stronger than 3-nitrophenol (pK_a 8.40) in which the –M effect cannot operate.

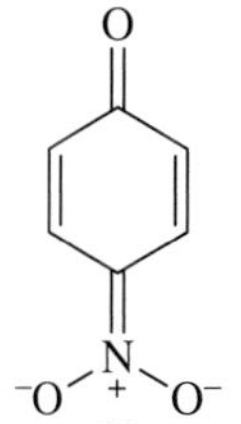

On the other hand, a methyl group exerts a weak +I effect and thus the methylphenols are slightly less acidic than phenol (*e.g.* 4-methylphenol, pK_a 10.14).

O
N
$^-$O + O$^-$
10

4.5 Reactions of the Hydroxy Group

4.5.1 Ester Formation and Fries Rearrangement

When treated with acid chlorides and acid anhydrides, phenols form esters. Under Friedel–Crafts conditions, phenolic esters undergo a **Fries rearrangement** in which the acyl group migrates to the 2- and 4-positions. Thus, treatment of the ester **11** with aluminium chloride in an inert solvent gives a mixture of 2- and 4-hydroxyacetophenones [(hydroxyphenyl)ethanones]; *C*-acylation has occurred (Scheme 4.7). The two isomers are separable and this is a useful method for the production of phenolic ketones. The mechanism remains uncertain, but it would appear that the acylium ion (RCO^+) is generated and that a Friedel–Crafts mechanism operates.

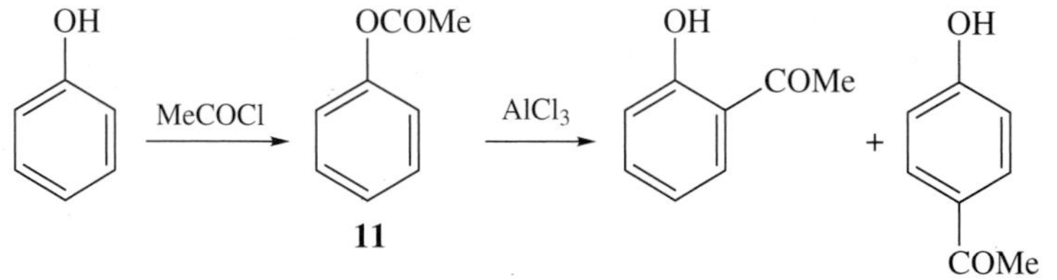

OH
MeCOCl
OCOMe
11
$AlCl_3$
OH
COMe
+
OH
COMe

Scheme 4.7

4.5.2 Ether Formation

When the sodium salt of a phenol is treated with an alkyl halide or an alkyl sulfate, *O*-alkylation occurs and an ether is formed, usually in good yield. Methyl ethers such as anisole (methoxybenzene) can also be formed in excellent yield by treatment of a phenol with diazomethane (Scheme 4.8).

Diazomethane, CH_2N_2, is an explosive, yellow gas which is a source of methylene, CH_2, and is an efficient methylating agent.

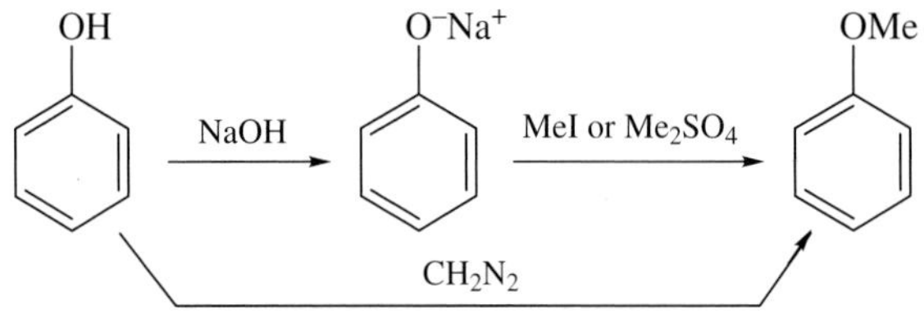

OH → (NaOH) → O^-Na^+ → (MeI or Me_2SO_4) → OMe; OH → (CH_2N_2) → OMe

Scheme 4.8

A reaction peculiar to allyl aryl ethers is their rearrangement to allylphenols when heated. In this **Claisen rearrangement**, the allyl group migrates to the 2-position. It is an example of a **pericyclic reaction** and proceeds through a cyclic, six-membered transition state (Scheme 4.9). The reaction has been investigated by labelling the γ-carbon atom with the isotope ^{14}C, marked with an asterisk in the scheme. The remote labelled carbon atom in the original allylphenol becomes the α-carbon attached to the ring in the product. If both *ortho* positions are occupied, then migration to the *para* position occurs in two stages and the label occupies the γ-position as in the starting material.

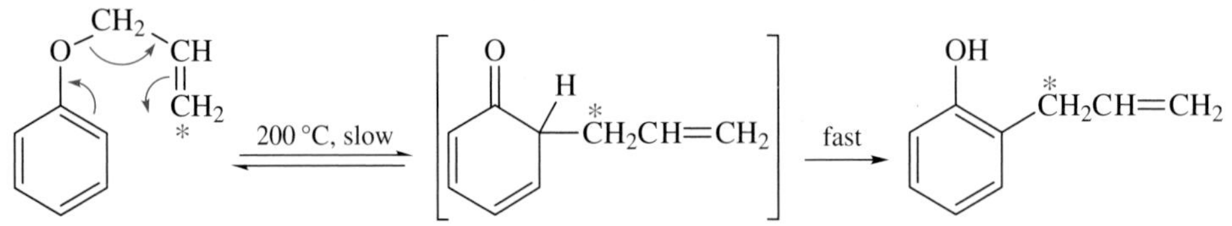

$O-CH_2-CH=\overset{*}{C}H_2$ ⇌ (200 °C, slow) [$\overset{*}{C}H_2CH=CH_2$ (cyclohexadienone, H)] → (fast) OH, $\overset{*}{C}H_2CH=CH_2$

Scheme 4.9

Worked Problem 4.2

Q Give a mechanistic explanation for the migration of the allyl group to the 4-position when the 2,6-dimethyl derivative of allylphenol is heated at 200 °C (Scheme 4.10).

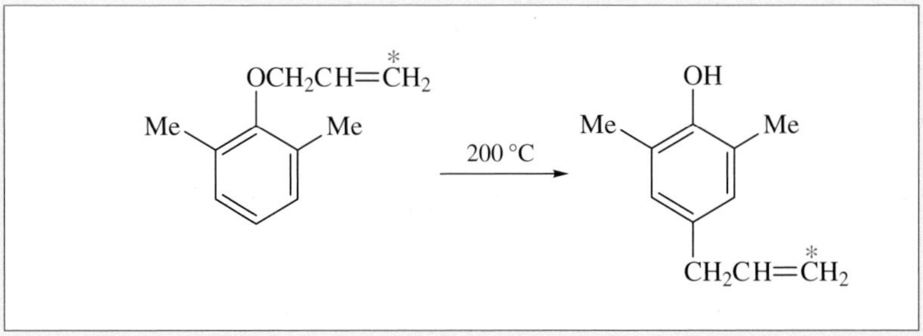

Scheme 4.10

A It has been shown with the aid of labelling experiments that the reaction proceeds in two stages, each involving a head-to-tail switch of the allyl group and each proceeding through a cyclic transition state (Scheme 4.11).

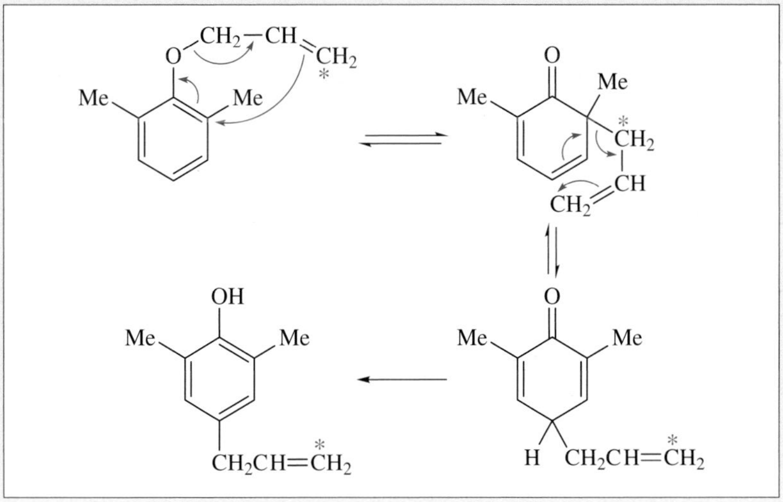

Scheme 4.11

4.6 Reactions of the Ring

4.6.1 Electrophilic Substitution

Phenols are highly activated towards electrophilic attack, which occurs readily at the 2- and 4-positions. For example, phenol reacts with bromine at room temperature in ethanol and in the absence of a catalyst to give 2,4,6-tribromophenol. Other electrophilic substitution reactions such as nitration, sulfonation, Friedel–Crafts, chlorination and nitrosation also proceed readily and hence care is needed to ensure multisubstitution does not occur. Protection of specific ring positions can also prevent unwanted substitution. Relatively mild conditions are usually employed.

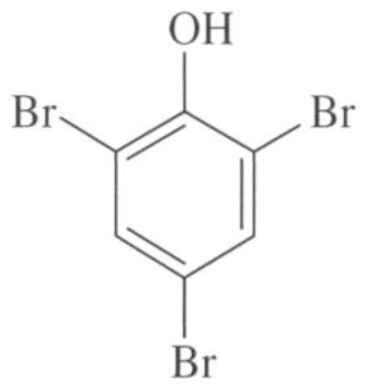

One synthesis of the analgesic paracetamol (acetaminophen, **12**) involves the nitrosation of phenol followed by reduction of the 4-nitrosophenol. Subsequent selective acetylation of the more reactive amino group completes the process:

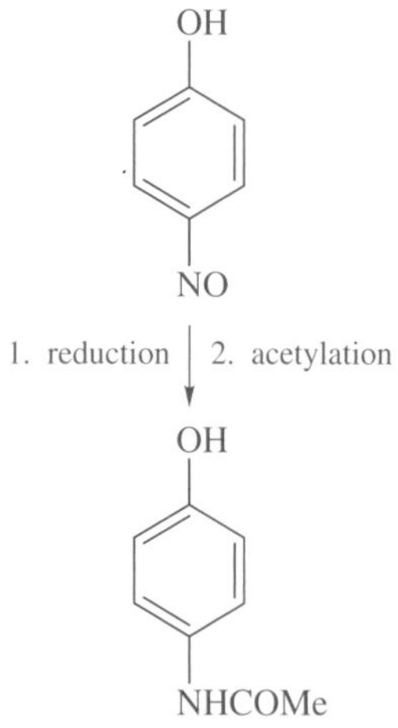

OH

NO

1. reduction | 2. acetylation

OH

NHCOMe
12

Dichlorocarbene is an example of a species that contains a divalent carbon atom. This carbon has two bond pairs of electrons and a lone pair and thus lacks an octet of electrons. Hence it can accept a pair of electrons, behaving as a Lewis acid or an electrophile. In one form **13** (a singlet carbene), dichlorocarbene can be pictured as a V-shaped species in which the carbon atom is sp^2 hybridized, with one orbital containing the lone pair and with an empty p orbital:

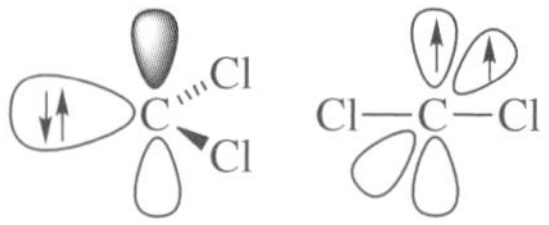

13 **14**

In the triplet state **14** the two electrons are unpaired in sp orbitals.

4.6.2 Reactions of the Phenoxide Ion

Under alkaline conditions, the phenoxide ion is formed, which is even more nucleophilic than phenol and hence more reactive. A number of C–C bond-forming reactions take place under these conditions.

An important reaction of phenols is the attack by weakly electrophilic arenediazonium salts in aqueous alkaline solution at below 5 °C to form azo dyes. This coupling reaction is discussed in Chapter 8.

The Reimer–Tiemann Reaction

Treatment of a phenol with chloroform (trichloromethane) in the presence of hydroxide ion results in the synthesis of a 2-hydroxybenzaldehyde through *C*-formylation. Dichlorocarbene, $:CCl_2$, is generated by the action of base on chloroform and this highly reactive electrophile then attacks the phenoxide. The mechanism of the **Reimer–Tiemann reaction** is given in Scheme 4.12.

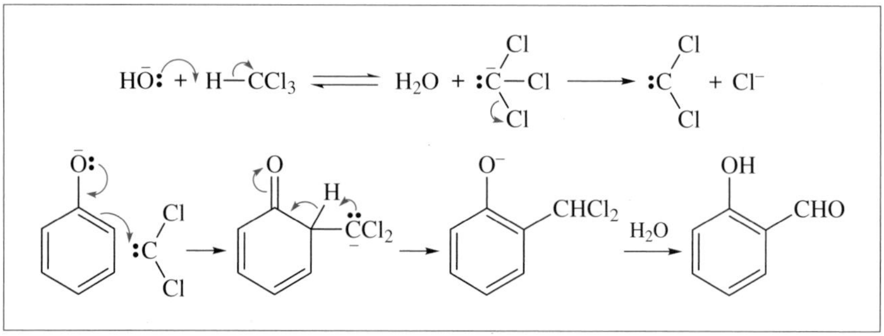

$$HO^- + H{-}CCl_3 \rightleftharpoons H_2O + {}^-:CCl_3 \longrightarrow :CCl_2 + Cl^-$$

Scheme 4.12

The Kolbe–Schmidt Reaction

The phenoxide ion is sufficiently nucleophilic to be attacked by carbon dioxide, providing a useful method for the introduction of a carboxylic acid group; *ortho* carboxylation takes place at 120–140 °C. The product of the **Kolbe-Schmidt reaction** on phenol is 2-hydroxybenzoic acid (salicylic acid) (Scheme 4.13).

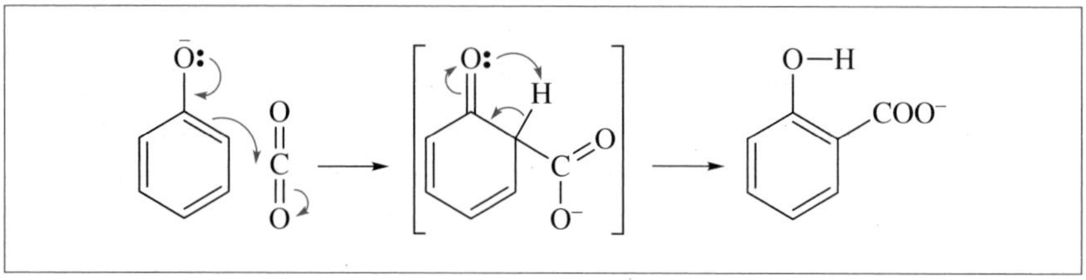

Scheme 4.13

With Formaldehyde

In aqueous alkaline solution, phenol reacts with formaldehyde (methanal) at low temperatures to form a mixture of 2- and 4-hydroxybenzyl alcohols. This **Lederer–Manasse reaction** is another example of electrophilic attack which results in the formation of a new C–C bond. The mechanism is illustrated in Scheme 4.14. These products readily lose water to form quinomethanes (methylenecyclohexadienones), which react with more phenoxide. This process is repeated over and over again to produce a cross-linked polymer or phenol–formaldehyde resin (*e.g.* Bakelite) in which the aromatic rings are linked to methylene bridges.

Reaction of 2,4,5-trichlorophenol (the antiseptic TCP) with HCHO yields hexachlorophene, a widely used germicide.

Note the ease with which the $CHCl_2$ group in the penultimate product in the Reimer–Tiemann reaction is hydrolysed. This behaviour is typical of a *gem*-dihalide unit.

Acetylation of the hydroxy group of salicylic acid gives aspirin, acetylsalicyclic acid.

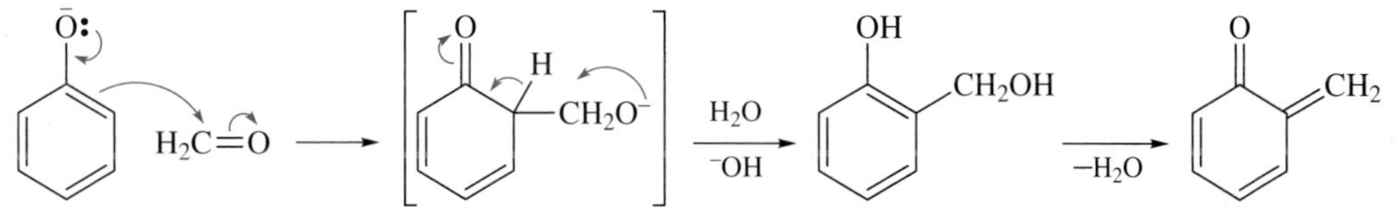

$H_2C{=}O$, CH_2O^-, H_2O / ^-OH, OH, CH_2OH, $-H_2O$, O, CH_2

o-Quinomethane

Scheme 4.14

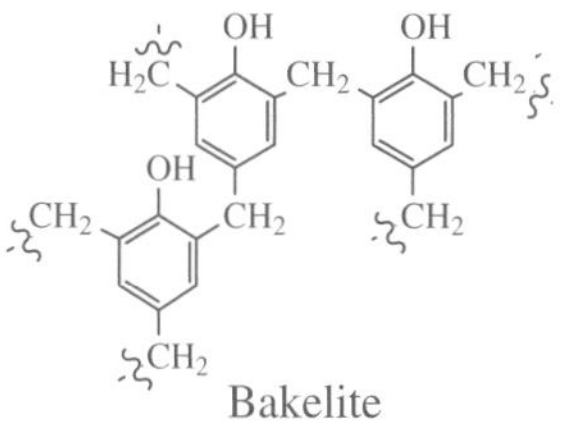

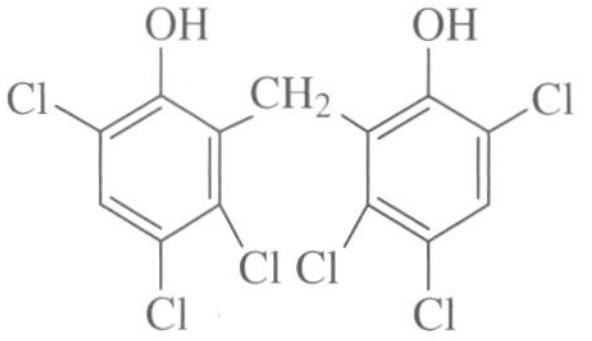

4.7 Dihydroxybenzenes

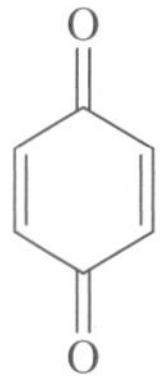

The dihydroxybenzenes or dihydric phenols **15–17** have trivial names as shown.

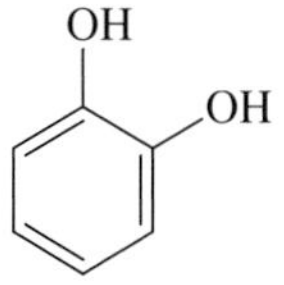

15 1,2-Dihydroxybenzene (catechol) 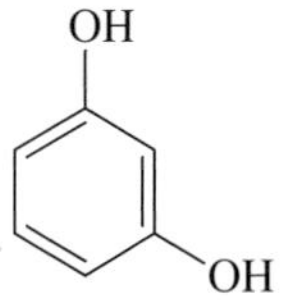 **16** 1,3-Dihydroxybenzene (resorcinol)  **17** 1,4-Dihydroxybenzene (hydroquinone)

Bakelite

Hexachlorophene

Benzo-1,4-quinone

1,2-Dihydroxybenzene may be prepared from 2-hydroxybenzaldehyde by the **Dakin reaction**, which involves oxidation in alkaline solution by hydrogen peroxide (Scheme 4.15). The reaction involves a 1,2-shift to an electron-deficient oxygen and is similar to the cumene process used to synthesize phenol (Section 4.2).

1,3-Dihydroxybenzene is prepared industrially by the alkali fusion of benzene-1,3-disulfonic acid. 1,4-Dihydroxybenzene is prepared in large quantities for use as a photographic developer, one process being by the oxidation of aniline with manganese dioxide [manganese(IV) oxide] in sulfuric acid to give benzo-1,4-quinone, which is then reduced to 1,4-dihydroxybenzene (hydroquinone, quinol).

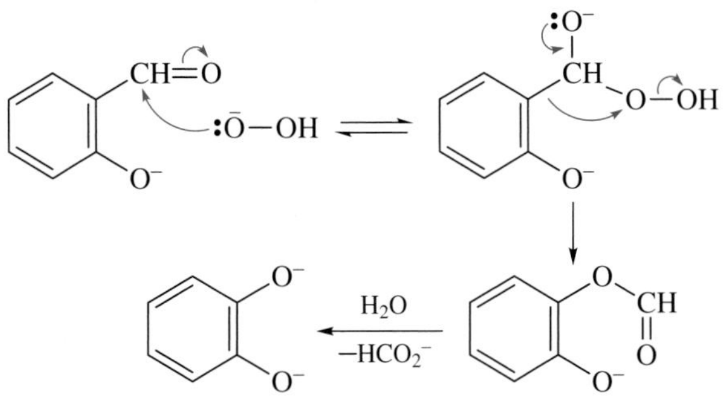

Scheme 4.15

Summary of Key Points

1. Industrially, the majority of phenol is produced by the oxidation of isopropylbenzene (cumene).

2. Phenols can be prepared from aromatic hydrocarbons by sulfonation and hydrolysis of the resulting sulfonic acid.

3. The hydroxy group is a strong electron donor and directs electrophilic attack to the *ortho* and *para* positions. Attack by electrophiles is very easy.

4. Phenols are less acidic than carboxylic acids, but more acidic than alcohols. Electron-withdrawing groups increase the acidity, but electron-donating groups reduce the acidity.

5. Phenols can be acylated and etherified.

Problems

4.1. Explain which is the more acidic phenol in each of the following pairs: (a) 3-cyanophenol and 4-cyanophenol; (b) 4-chlorophenol and 4-methoxyphenol; (c) 2,4-dinitrophenol and 3,5-dinitrophenol; (d) 4-hydroxybenzaldehyde and 4-nitrophenol.

4.2. Suggest the main organic product formed in the following reactions: (a) phenol + toluene-4-sulfonyl chloride in pyridine; (b) 2,4,5-trichlorophenol + chloroacetic acid + NaOH; (c) heating pent-2-enyl phenyl ether at 200 °C; (d) 4-methylphenol + Ac_2O + $AlCl_3$ + heat; (e) anisole + succinic anhydride + $AlCl_3$ at 0 °C.

4.3. How would you prepare the following compounds from the named starting material: (a) 1-methoxy-3-nitrobenzene from benzenesulfonic acid; (b) methyl 2-hydroxybenzoate from phenol; (c) 2-bromo-4-methylphenol from toluene.

4.4. Explain why 2,4,6-trinitrophenol reacts with aqueous sodium hydrogen carbonate solution but phenol does not.

5
Aromatic Acids

Aims

By the end of this chapter you should understand:

- How carboxylic and sulfonic acid groups are introduced into aromatic molecules
- The nature of the acidity of these groups
- The effect of the acidic groups on ring reactivity
- The reactions of the acidic groups

5.1 Introduction

Both the sulfonic acid ($-SO_3H$) and the carboxylic acid ($-CO_2H$) groups are encountered in aromatic molecules. Introduction of one sulfonic acid group into the benzene ring gives **benzenesulfonic acid** (**1**), derivatives of which are named as the substituted benzenesulfonic acid. The corresponding carboxylic acid is **benzoic acid** (**2**).

Benzenedicarboxylic acids have trivial names. Benzene-1,4-dicarboxylic acid (terephthalic acid, **3**) is used in the manufacture of commercially important polyesters. Esters of benzene-1,2-dicarboxylic acid (phthalic acid) are used for plasticizing polymers.

SO_3H CO_2H

1 **2**

CO_2H

CO_2H

3

5.2 Introduction of Acidic Groups

The methods of introducing the two groups are quite different. Sulfonic acids are usually obtained by direct electrophilic substitution, whilst carboxylic acids are produced through the conversion of another functional group.

5.2.1 Introduction of the Sulfonic Acid Group

Benzene reacts slowly with hot sulfuric acid to produce benzenesulfonic acid. The attacking electrophile, the cation **4**, is generated by the self-protonation of sulfuric acid and reacts with the benzene ring in the normal manner (Scheme 5.1).

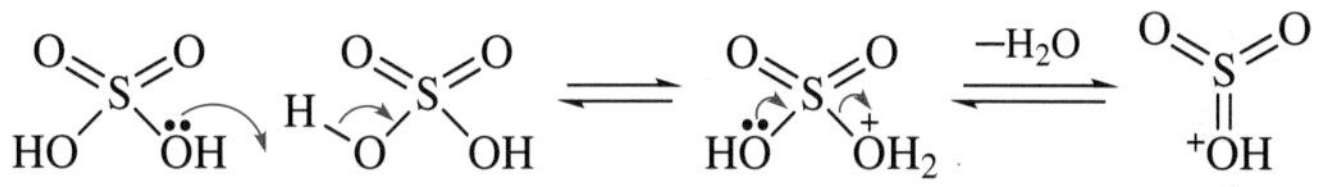

Scheme 5.1

Dissolving sulfur trioxide, SO_3, in sulfuric acid forms the same cation and these solutions, known as oleum or fuming sulfuric acid, readily sulfonate benzene and even less reactive aromatic systems.

65% oleum (65% w/v solution of SO_3 in sulfuric acid) is a very powerful reagent that reacts extremely violently with water and has to be handled with great care.

There is some evidence from kinetic studies that the electron-deficient and therefore electrophilic sulfur trioxide is itself the attacking species, when the mechanistic pathway follows that illustrated in Scheme 5.2.

$$2H_2SO_4 \rightleftharpoons H_3O^+ + HSO_4^- + SO_3$$

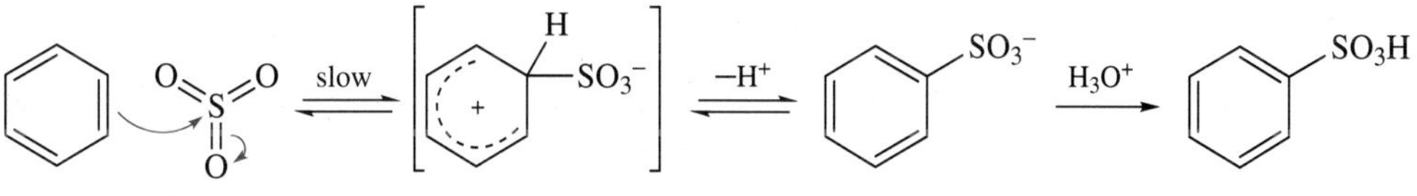

Scheme 5.2

Chlorosulfonic acid also effects direct sulfonation, although when used in excess the product is the sulfonyl chloride; subsequent hydrolysis leads to the acid (Scheme 5.3).

Sulfonation is a reversible reaction. A combination of sulfonation and desulfonation is a useful means of directing electrophilic attack to specific positions, so providing a route to compounds otherwise difficult to prepare.

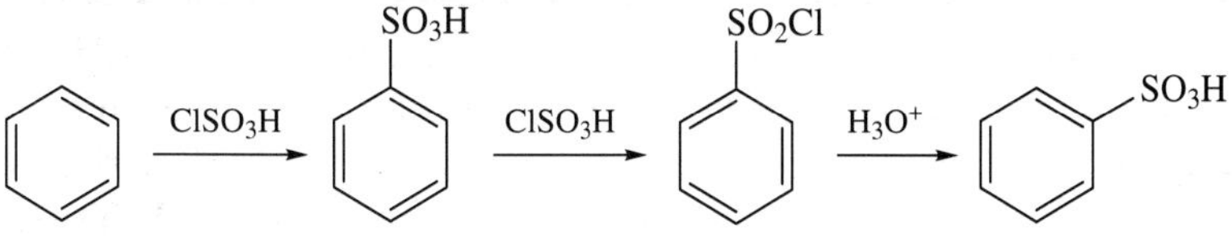

Scheme 5.3

5.2.2 Introduction of the Carboxylic Acid Group

Oxidative Methods

A variety of oxidizing agents, including potassium permanganate [potassium manganate(VII)], chromium trioxide [chromium(VI) oxide] in sulfuric acid, potassium dichromate and hydrogen peroxide, convert alcohols, aldehydes, alkyl and halogenated alkyl groups to carboxylic acids (Scheme 5.4). For instance, benzaldehyde is readily oxidized to benzoic acid in good yield by potassium permanganate.

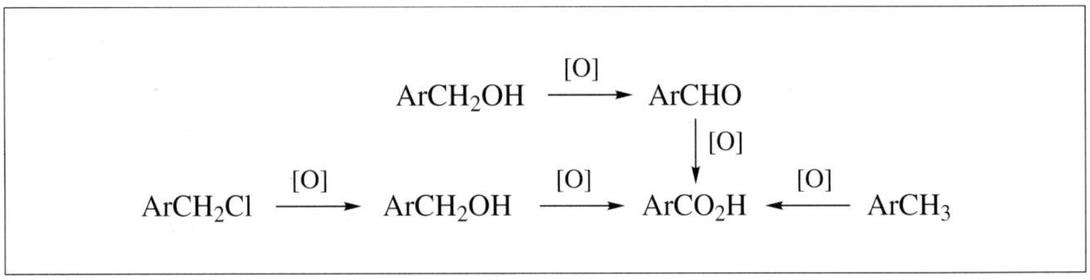

$$ArCH_2OH \xrightarrow{[O]} ArCHO$$

$$ArCHO \xrightarrow{[O]} ArCO_2H$$

$$ArCH_2Cl \xrightarrow{[O]} ArCH_2OH \xrightarrow{[O]} ArCO_2H \xleftarrow{[O]} ArCH_3$$

Scheme 5.4

Hydrolytic Methods

Hydrolysis of acid chlorides, acid anhydrides, esters and carboxamides leads to the carboxylic acid, although these compounds are often derived from a carboxylic acid group in the first place (Scheme 5.5). Nitriles are usually derived from amines *via* diazotization and reaction with copper(I) cyanide (see Chapter 8) and so the hydrolysis of a nitrile group is of more value. In all cases, alkaline hydrolysis gives the salt of the acid, from which the free acid is obtained by addition of mineral acid.

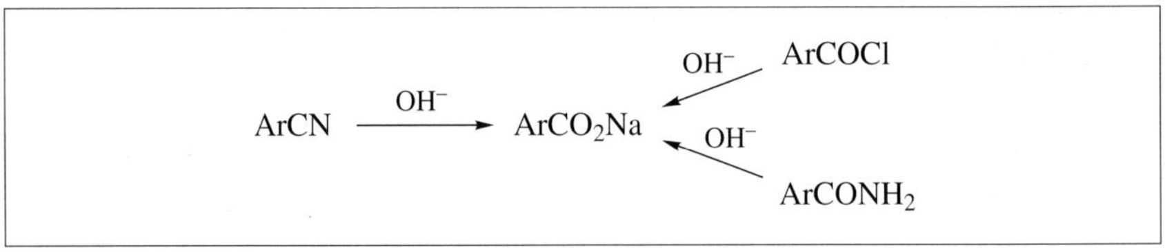

$$ArCN \xrightarrow{OH^-} ArCO_2Na$$

$$ArCOCl \xrightarrow{OH^-} ArCO_2Na \xleftarrow{OH^-} ArCONH_2$$

Scheme 5.5

Miscellaneous Methods

Aromatic carboxylic acids are readily prepared from aryl halides by conversion to the Grignard reagent or aryllithium compound and subsequent reaction with carbon dioxide (see Chapter 10).

Benzoic acid is prepared industrially by the oxidation of toluene using air at 170 °C over a catalyst of cobalt and manganese acetate. An alternative route involves the hydrolysis of (trichloromethyl)benzene using aqueous calcium hydroxide in the presence of iron powder as catalyst; $PhCCl_3$ is prepared by chlorination of toluene in the presence of light.

5.3 Reactions of Aromatic Acids

5.3.1 Reactions of the Acid Group

Sulfonic acids and carboxylic acids can be converted into their acid chlorides by treatment with phosphorus pentachloride or phosphorus oxychloride. Thionyl chloride, $SOCl_2$, is effective for the synthesis of acyl chlorides, and sulfonyl chlorides can be prepared directly from the aromatic compound by reaction with an excess of chlorosulfonic acid. The acid chlorides are efficient Friedel–Crafts acylating agents, yielding sulfones, $ArSO_2Ph$, and ketones, ArCOPh. They react readily with alcohols to form esters and with ammonia or amines to form sulfonamides or carboxamides (Scheme 5.6). The reaction of benzoyl chloride, PhCOCl, with amines in the presence of aqueous sodium hydroxide is known as the **Schotten–Baumann reaction** and has been used to characterize amines.

Alkylated aromatic sulfonic acids are used as detergents. Sulfonamides possess antibacterial properties.

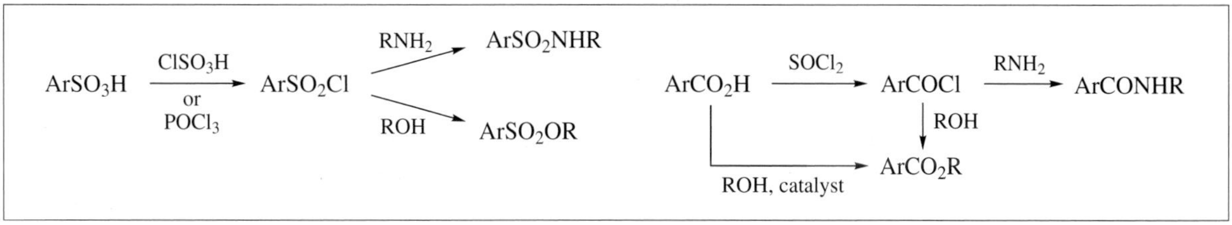

$$ArSO_3H \xrightarrow[\text{or } POCl_3]{ClSO_3H} ArSO_2Cl \begin{cases} \xrightarrow{RNH_2} ArSO_2NHR \\ \xrightarrow{ROH} ArSO_2OR \end{cases}$$

$$ArCO_2H \xrightarrow{SOCl_2} ArCOCl \xrightarrow{RNH_2} ArCONHR$$

$$ArCOCl \xrightarrow{ROH} ArCO_2R \qquad ArCO_2H \xrightarrow{\text{ROH, catalyst}} ArCO_2R$$

Scheme 5.6

Worked Problem 5.1

Q Acetyl chloride reacts almost explosively with cold water, whereas benzoyl chloride reacts only slowly even with dilute sodium hydroxide. Explain.

A The mechanism of hydrolysis is the same for both compounds, involving nucleophilic attack by water or OH^- at the electron-deficient carbon atom of the acyl chloride function. However, the extent of this deficiency is different in the two acid chlorides. In acetyl chloride, electron withdrawal by both the carbonyl group and the chlorine atom make the carbon significantly electropositive, promoting the sequence shown in Scheme 5.7.

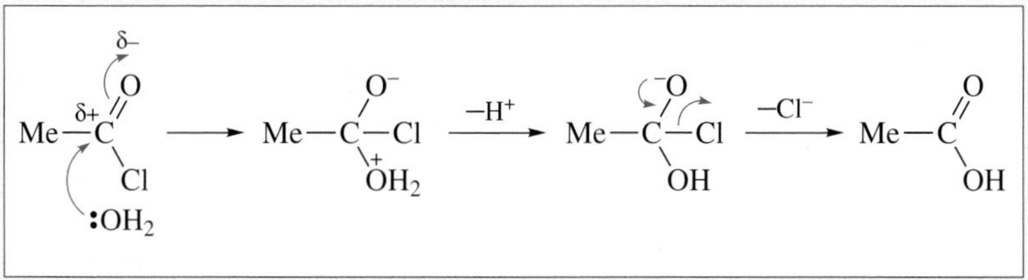

$$Me\!-\!C^{\delta+}(=O^{\delta-})Cl + {:}OH_2 \longrightarrow Me\!-\!C(O^-)(Cl)(\overset{+}{O}H_2) \xrightarrow{-H^+} Me\!-\!C(O^-)(Cl)(OH) \xrightarrow{-Cl^-} Me\!-\!C(=O)OH$$

Scheme 5.7

On the other hand, in benzoyl chloride the aromatic ring shares in this electron withdrawal. The resistance to hydrolysis shown by benzoyl chloride follows from the reduced electron deficiency of the carbon atom that results from this resonance interaction (Scheme 5.8).

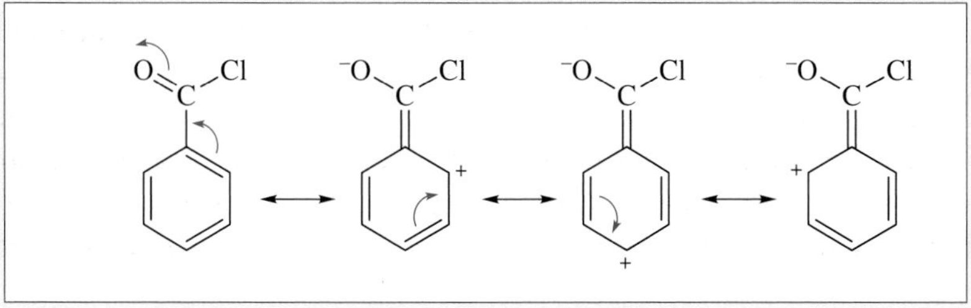

$$PhC(=O)Cl \longleftrightarrow {}^-O\!-\!C(Cl)\!=\!C_6H_5^+\,(ortho^+) \longleftrightarrow {}^-O\!-\!C(Cl)\!=\!C_6H_5^+\,(para^+) \longleftrightarrow {}^-O\!-\!C(Cl)\!=\!C_6H_5^+\,(ortho'^+)$$

Scheme 5.8

Dibenzoyl peroxide, $(PhCO)_2O_2$, is a valuable free radical generator. Peroxy acids, RCO_3H, which also contain a peroxy unit, O–O, are useful oxidizing agents (*e.g.* in the synthesis of epoxides).

Benzoic acid anhydride is formed by the reaction of sodium benzoate with benzoyl chloride. Dibenzoyl peroxide and peroxybenzoic acid can be prepared from the acid chloride by reaction with hydrogen peroxide (Scheme 5.9).

$$PhCO_2Na + PhCOCl \longrightarrow (PhCO)_2O + NaCl$$

Benzoic acid anhydride

$$2PhCOCl + 2NaOH + H_2O_2 \longrightarrow (PhCO)_2O_2 \xrightarrow{MeONa} PhCO_3Na + PhCO_2Me$$

Dibenzoyl peroxide; Sodium perbenzoate

Scheme 5.9

5.3.2 Displacement Reactions of the Sulfonic Acid Group

The sulfonation reaction is reversible and benzenesulfonic acid may be desulfonated by treatment with dilute acid at 150 °C. The group can be displaced by fusion of its salt with sodamide to give the corresponding amine, with sodium hydroxide to give the phenol, sodium cyanide to give the nitrile, and potassium hydrogen sulfide to give the benzenethiol (Scheme 5.10).

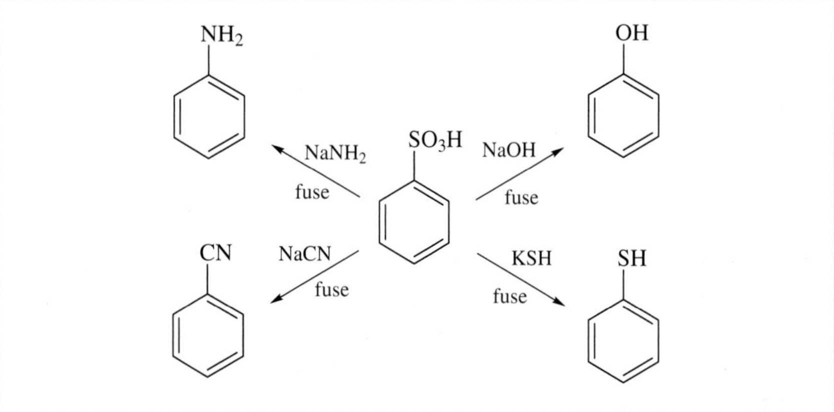

Scheme 5.10

The sulfonic acid group can undergo *ipso* substitution (see Chapter 2) with nitric acid, resulting in the introduction of a nitro group.

5.3.3 Reactions of the Ring

Both the sulfonic acid group and the carboxylic acid group are deactivating and *meta* directing. Thus further electrophilic attack requires relatively forcing conditions.

5.4 Acidity of Aromatic Acids

Aromatic sulfonic acids are strong acids, of similar strength to sulfuric acid. *p*-Toluenesulfonic acid (4-TsOH) is used as an acid catalyst in various reactions.

Benzoic acids are weaker than sulfonic acids, benzoic acid itself having a pK_a of 4.17.

The effect of substituents on ionization has already been discussed in Chapter 2, where the Hammett substituent constant was described, and in Chapter 4 in connection with the acidity of phenols. These effects influence the relative acidity of benzoic acid derivatives in a similar way.

Any feature that stabilizes the anion relative to the parent acid will force the equilibrium below towards the anion and thus increase the acidity:

$$RCOOH + H_2O \rightleftharpoons RCOO^- + H_3O^+$$

The carboxyl group withdraws electron density from the aromatic ring, which implies that the ring donates electrons to the carboxyl group as illustrated by structures **5** and **6** in Scheme 5.11. Such behaviour is acid weakening and benzoic acid (pK_a 4.17) is weaker than formic acid (methanoic acid, HCO_2H; pK_a 3.75). Note that electron donation by the phenyl group is apparently less than that of a methyl group, since acetic acid (ethanoic acid; pK_a 4.76) is a weaker acid than benzoic acid.

Dispersal of the positive charge from the carboxyl carbon atom into the ring also accounts for the lack of carbonyl group properties of carboxylic acids.

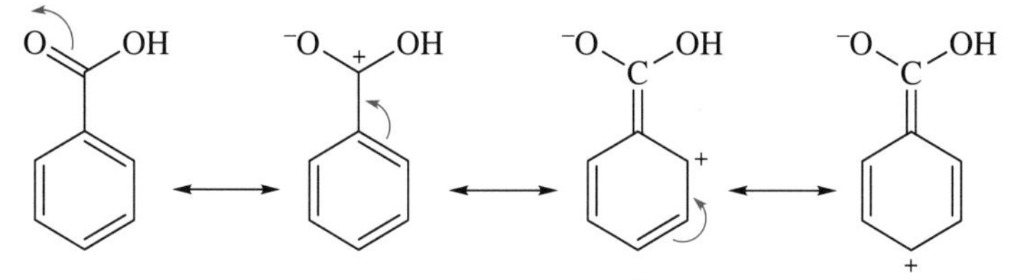

Scheme 5.11

Substituents in the 3- and 4-positions of the phenyl ring influence the acidity of benzoic acids in accordance with their ability to donate or withdraw electron density from the carboxyl function. Electron-donating substituents decrease the acidity through their +I and +M effects; this effect is quite small, as can be seen from the data in Table 5.1.

Groups that reduce the electron density at the carbon atom to which the carboxylic acid group is attached assist proton release and increase the acidity. In cases where the substituent can exert both –I and –M effects, the 4-isomer is a stronger acid than the 3-substituted compound, for which only inductive withdrawal can operate. Thus, 4-nitrobenzoic acid is a stronger acid than 3-nitrobenzoic acid, but both are stronger than benzoic acid. The converse applies for a –I/+M substituents such as methoxy, since the mesomeric effect decreases the acidity of 4-methoxybenzoic acid (pK_a 4.47), whereas the electron-withdrawing

Table 5.1 Acidity of some substituted benzoic acids

Substituent	pK_a		
	ortho	*meta*	*para*
H	4.17	4.17	4.17
Me	3.91	4.27	4.37
OMe	4.09	4.09	4.47
OH	2.98	4.08	4.58
NO_2	2.17	3.49	3.43
Cl	2.94	3.83	3.98

inductive effect alone operates in 3-methoxybenzoic acid, making it a slightly stronger acid (pK_a 4.09) than benzoic acid.

It is less easy to predict the influence that 2-substituents have on the acidity of benzoic acids because other effects may operate as a result of the closeness of the two functions. Intramolecular hydrogen bonding between the carboxy carbonyl group and a 2-substituent markedly increases the acidity through the greater stabilization of the anion. For example, 2-hydroxybenzoic acid (salicylic acid, pK_a 2.98) is a stronger acid than the two isomeric hydroxybenzoic acids because of the efficient stabilization of the anion **7**. An *ortho* substituent may also exert a steric effect that reduces the co-planarity of the carboxyl group with the aromatic ring, thereby decreasing electron donation by the ring and so increasing the acidity.

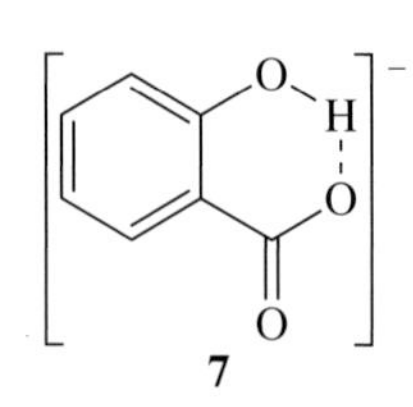

In summary, electron-withdrawing groups increase acidity and electron-donating systems decrease acidity when in the 3- and 4-positions. 2-Substituents generally increase acidity.

5.5 Compounds with More Than One Acidic Group

The benzenedisulfonic acids are of little interest, except that benzene-1,3-disulfonic acid is a source of 1,3-dihydroxybenzene (see Chapter 4). The benzene dicarboxylic acids are more important. Benzene-1,2-dicarboxylic acid (phthalic acid, **8**) can be converted into phthalic anhydride (**9**), which is a typical acid anhydride, reacting with amines and alcohols and also taking part in Friedel–Crafts reactions. Phthalimide (**10**), produced by reaction of the anhydride with ammonia, is weakly acidic and forms a potassium salt with ethanolic potassium hydroxide.

In **Gabriel's synthesis** of primary amines, potassium phthalimide is reacted with an alkyl halide and the resulting *N*-alkylphthalimide is hydrolysed to release the amine and phthalic acid (Scheme 5.12). This route is also used in the synthesis of amino acids.

1,4-Benzenedicarboxylic acid is used as its dimethyl ester in the syn-

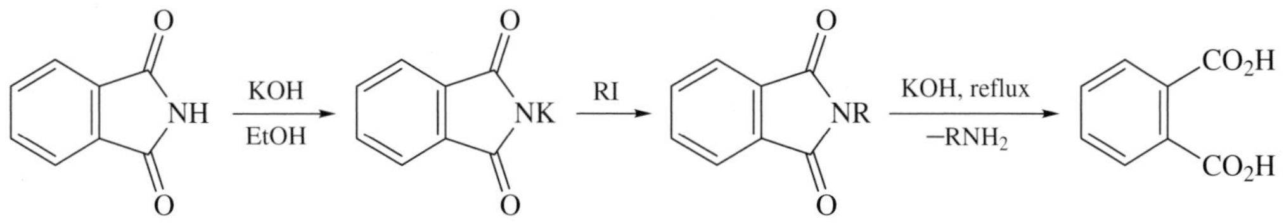

Phthalimide (NH) $\xrightarrow[\text{EtOH}]{\text{KOH}}$ potassium phthalimide (NK) $\xrightarrow{\text{RI}}$ N-alkylphthalimide (NR) $\xrightarrow[-\text{RNH}_2]{\text{KOH, reflux}}$ phthalic acid (CO_2H, CO_2H)

Scheme 5.12

thesis of poly(ethylene terephthalate) (PET), two commercial names of which are Terylene and Dacron, used in the manufacture of clothing, carpets and drinks bottles.

5.6 Side-chain Acids

Phenylacetic acid (phenylethanoic acid, **11**), the simplest example of this type of compound, is prepared from benzyl chloride by the S_N displacement of chloride by cyanide ion and subsequent hydrolysis to the acid (Scheme 5.13).

$$PhCH_2Cl \xrightarrow{\text{KCN}} PhCH_2CN \xrightarrow[\text{reflux}]{\text{aq. HCl}} \underset{\mathbf{11}}{PhCH_2CO_2H}$$

Scheme 5.13

Unsaturated side-chains are present in the cinnamic acids, some of which occur naturally in the *trans* form **12**. Syntheses of the acids and their esters by the Claisen condensation and the Perkin and Knoevenagel reactions are discussed in Chapter 6.

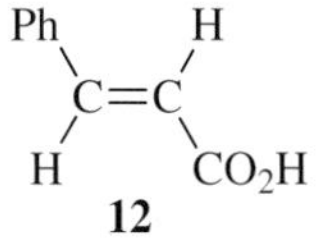

2-Hydroxycinnamic acids spontaneously cyclize to give a coumarin:

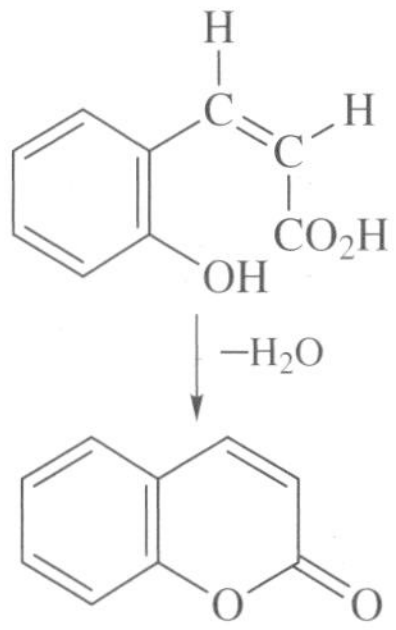

2-HOC$_6$H$_4$CH=CHCO$_2$H $\xrightarrow{-H_2O}$ coumarin

Some coumarins possess anticoagulant properties (warfarin is a widely used example) and other derivatives are used as fluorescent brightening agents. A number of laser dyes are based on coumarin. Cinnamic acid esters are commonly used in sun-tan lotions to absorb UV light.

$$\underset{\mathbf{12}}{(Ph)(H)C{=}C(H)(CO_2H)}$$

3-Phenylpropynoic acid (**13**) can be prepared from cinnamic acid esters by the addition of bromine to the double bond followed by dehydrobromination with alcoholic potassium hydroxide (Scheme 5.14).

$$PhCH{=}CHCO_2Et \xrightarrow{Br_2} PhCHBrCHBrCO_2Et \xrightarrow[\text{2. H}^+]{\text{1. KOH, EtOH}} \underset{\mathbf{13}}{PhC{\equiv}CCO_2H}$$

Scheme 5.14

Summary of Key Points

1. Benzenesulfonic acids are prepared by direct sulfonation of aromatic hydrocarbons.

2. Aromatic carboxylic acids can be prepared by oxidation of an alkyl side-chain and by the reaction of an aryl Grignard reagent with CO_2.

3. Both carboxylic and sulfonic acids can be converted into esters, anhydrides and amides.

4. The sulfonic acid group can be displaced by a variety of nucleophiles.

5. Sulfonic acids are stronger acids than the corresponding carboxylic acids. Electron-withdrawing groups increase the acidity of both types, but electron-releasing groups decrease the acidity.

Problems

5.1. Using curly arrows, show: (a) how the cyclization of 2-hydroxycinnamic acid to coumarin occurs; (b) how the alkaline hydrolysis of benzonitrile to sodium benzoate takes place.

5.2. Arrange the following compounds in order of increasing acid strength, giving reasons for your chosen order: 4-acetylbenzoic acid, 4-hydroxybenzoic acid, 4-cyanobenzoic acid, benzene-1,4-dicarboxylic acid.

5.3. Devise laboratory synthetic routes from toluene to (a) (4-nitrophenyl)acetic acid; (b) 3-methylbenzoic acid; (c) 4-sulfobenzoic acid; (d) 2,4-dibromobenzamide.

6 Aromatic Aldehydes, Ketones and Alcohols

Aims

By the end of this chapter you should understand:

- How aromatic aldehydes, ketones and alcohols can be synthesized
- Their reactions at both ring carbon atoms and the carbonyl group
- The generation of carbanions and their reactions with aromatic aldehydes and ketones

6.1 Introduction

Aromatic aldehydes and ketones show the usual reactions associated with a carbonyl group, but they display further reactions arising from the influence of the aromatic environment. This chapter describes synthetic routes to and the chemistries of benzaldehyde (**1**) and acetophenone (phenylethanone, **2**) and their derivatives.

In aromatic alcohols, of which the simplest example is benzyl alcohol (phenylmethanol, **3**), the hydroxyl group is present in an aliphatic side chain. Hence they are best regarded as aryl-substituted alcohols. Their properties are significantly different from those of phenols, but are typical of alcohols.

CHO COMe

1 **2**

CH_2OH

3

6.2 Aromatic Alcohols

6.2.1 Synthesis of Aromatic Alcohols

Benzyl alcohol (**3**) can be synthesized by the hydrolysis of (chloromethyl)benzene (benzyl chloride) (see Chapter 9) and by the reduction of benzaldehyde with sodium borohydride [sodium tetrahydridoborate(III)] (Scheme 6.1).

$$PhCH_2Cl \xrightarrow{NaOH} PhCH_2OH \xleftarrow{NaBH_4} PhCHO$$

Scheme 6.1

Reduction of the carbonyl function in acetophenones using sodium in ethanol also produces an alcohol. More complex aromatic alcohols may be prepared from carbonyl compounds by reaction with Grignard reagents followed by hydrolysis (see Chapter 10).

6.2.2 Reactions of Aromatic Alcohols

The typical reactions of the alcohol group include their conversion to ethers and esters by reaction with alkyl halides and with acid chlorides or anhydrides, respectively (Scheme 6.2). The benzyl ether group is readily cleaved by hydrogenolysis and is often used as a protecting group for alcohols. Primary alcohols are oxidized initially to the aldehyde and then to the carboxylic acid.

Hydrogenolysis means the removal of a functional group and its replacement by hydrogen. Here, the benzyl–oxygen bond is broken, a reaction that has proved important in peptide synthesis. Hydrogenation refers to the addition of hydrogen to, for example, a C=C bond.

$$PhCH_2OCH_2Ph \xleftarrow[2.\ PhCH_2Cl]{1.\ Na} PhCH_2OH \xrightarrow{[O]} PhCHO \xrightarrow{[O]} PhCO_2H$$

$$PhCH_2OH \xrightarrow{AcCl} PhCH_2OAc$$

Scheme 6.2

The CH_2OH group is *ortho*/*para* directing towards electrophilic attack. Nitration and sulfonation are possible, but care must be taken to avoid interaction with the hydroxyl group. It is sometimes preferable to carry out the electrophilic substitution reaction on the appropriate benzyl halide and then to hydrolyse the product to the substituted alcohol.

In order to prevent a functional group taking part in an unwanted reaction, the group is first reacted with a reagent which protects the functional group from the unwanted reaction. The desirable features of a protecting group are:

- It can easily be attached and subsequently removed under conditions that do not harm other functional groups present in the molecule.
- It makes the functional group inert to attack by reagents that would otherwise attack it.

6.3 Aromatic Aldehydes

6.3.1 Introduction of the Aldehyde Group

Benzaldehyde is prepared by the hydrolysis of (dichloromethyl)benzene (benzal chloride) in either aqueous acid or aqueous alkali and by the oxidation of toluene with chromium trioxide in acetic anhydride (Scheme 6.3). In the latter synthesis, as the benzaldehyde is formed, it is converted into its diacetate by the acetic anhydride, so preventing further oxidation; subsequent hydrolysis generates the aldehyde group. The benzaldehyde has thus been protected from oxidation. Benzyl alcohol can also be oxidized to benzaldehyde using chromium trioxide in acetic anhydride.

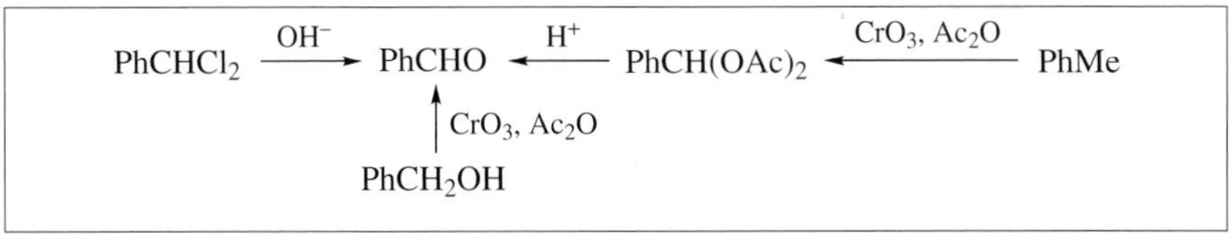

$$PhCHCl_2 \xrightarrow{OH^-} PhCHO \xleftarrow{H^+} PhCH(OAc)_2 \xleftarrow{CrO_3,\ Ac_2O} PhMe$$

$$PhCH_2OH \xrightarrow{CrO_3,\ Ac_2O} PhCHO$$

Scheme 6.3

There are several methods for the direct introduction of an aldehyde group into an aromatic compound. In the **Vilsmeier–Haack reaction**, activated aromatic systems such as aryl ethers and dialkylanilines are formylated by a mixture of dimethylformamide, $HCONMe_2$, and phosphorus oxychloride, $POCl_3$, (Scheme 6.4). The process involves electrophilic attack by a chloroiminium ion, $Me_2\overset{+}{N}{=}CHCl$, formed by interaction of dimethylformamide and phosphorus oxychloride. Hydrolysis of the dimethyl imine completes the synthesis.

Formylation is the direct introduction of the formyl group, CH=O, into a molecule.

In this chapter in particular, many reactions will be met which are named after their discoverers. Although it can be helpful to remember the names, it is far more important to understand why the reactions occur and the mechanisms by which they proceed.

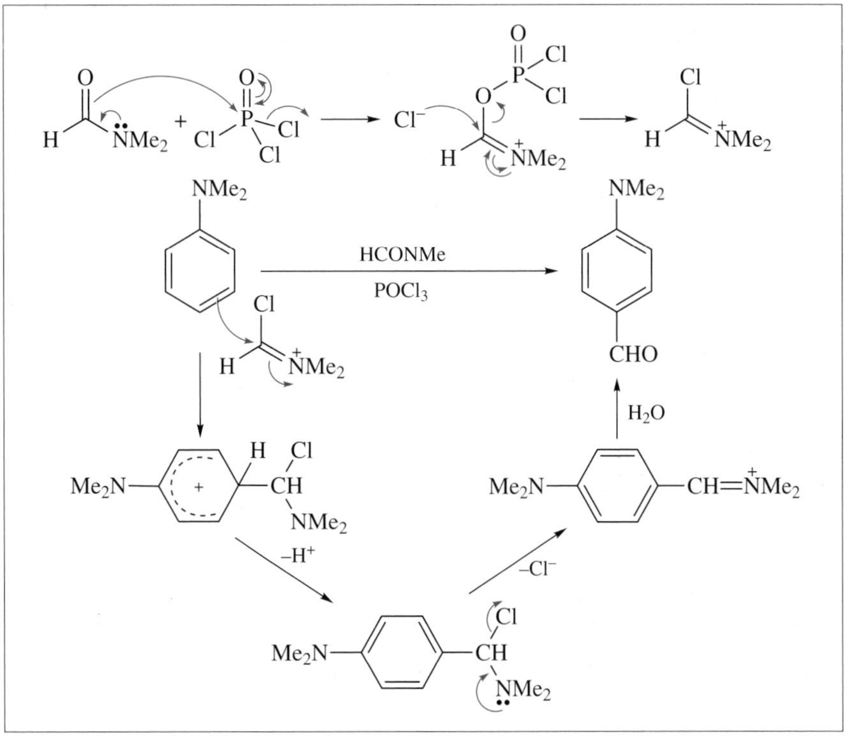

Scheme 6.4

The **Gattermann–Koch reaction**, in which carbon monoxide and hydrogen chloride are bubbled through a solution containing benzene and aluminium chloride, is a further example of direct formylation (Scheme 6.5). The formyl cation, $H\overset{+}{C}{=}O$, is thought to be the attacking electrophile, though it is probably complexed to Al.

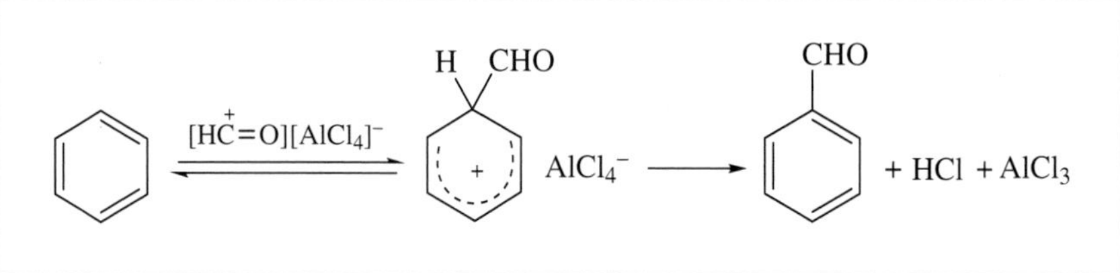

Scheme 6.5

In a related reaction, the **Gattermann aldehyde synthesis**, the carbon monoxide of the previous reaction is replaced by hydrogen cyanide (Scheme 6.6). This reaction gives poor yields with benzene itself, but is successful with activated species such as aryl ethers and phenols. The reaction proceeds *via* an aryl imine and the mechanism is not dissimilar to that of the Vilsmeier–Haack reaction.

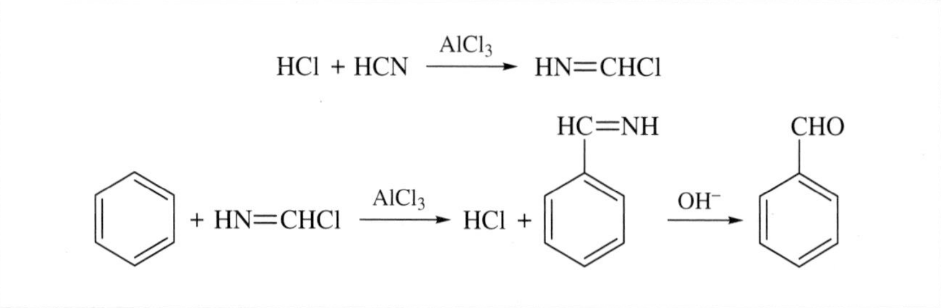

Scheme 6.6

The synthesis of aldehydes by the **Sommelet reaction**, the **Rosenmund reduction** and the **Stephens reaction** all involve the conversion of a group already present in the molecule. The Rosenmund reduction (Scheme 6.7) is the catalytic hydrogenation of a benzoyl chloride in the presence of a catalyst poison, quinoline/sulfur, which prevents over-reduction to the alcohol. In the Stephens reaction (Scheme 6.7), a nitrile group is reduced by tin(II) chloride and hydrochloric acid to an imine salt, which is then hydrolysed.

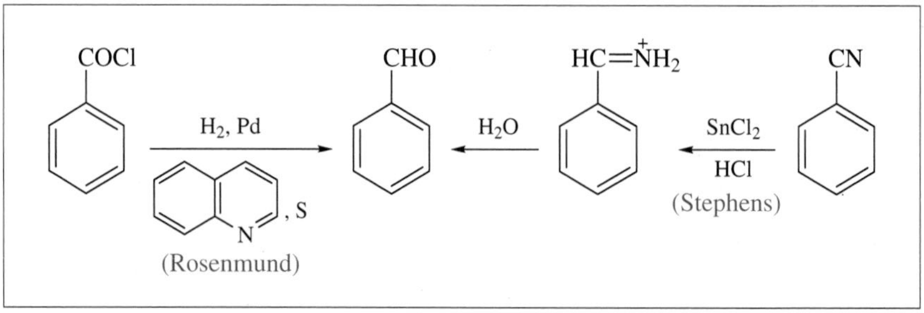

Scheme 6.7

The Sommelet reaction (Scheme 6.8) involves refluxing (chloromethyl)-benzene in aqueous ethanolic solution with hexamethylenetetramine followed by acidification.

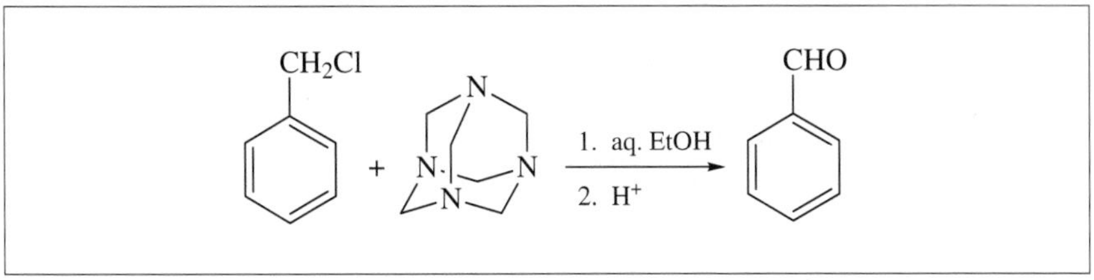

CH_2Cl + (hexamethylenetetramine) $\xrightarrow{\text{1. aq. EtOH; 2. } H^+}$ CHO

Scheme 6.8

6.3.2 Reactions of Aldehydes

Reactions at Ring Carbon

The aldehyde group deactivates the ring and is *meta* directing. There are few useful examples, since not only is electrophilic attack more difficult than for benzene, but also the aldehyde group is prone to oxidation during the attack. Substituted benzaldehydes are therefore usually synthesized by functional group transformations or by direct formylation.

Reactions of the Aldehyde Group

The carbonyl group is a reactive function and, although aromatic aldehydes are somewhat less reactive than their aliphatic counterparts, benzaldehydes have an extensive chemistry. Many reactions replicate those of aliphatic aldehydes, but are mentioned here for completeness. Thus, oxidation of the carbonyl group leads to carboxylic acids and reduction gives alcohols. The aldehyde group reacts with a range of N-nucleophiles (Scheme 6.9). Imines (Schiff bases) are formed with amines and hydrazones with hydrazines. Semicarbazide gives semicarbazones and hydroxylamine forms oximes.

Hydrazones and oximes are of value in the synthesis of heterocycles. For example, the acid-catalysed cyclization of phenylhydrazones gives indoles. 1,3-Diketones react with hydroxylamine to give isoxazoles.

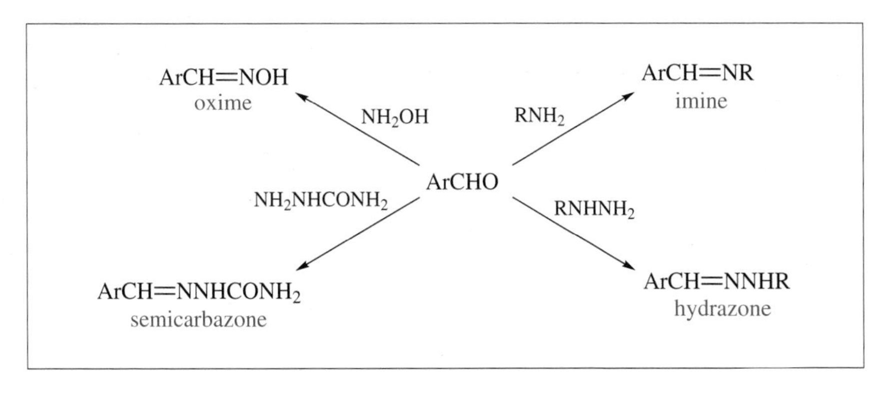

Scheme 6.9

The mechanism proceeds by nucleophilic attack of the nitrogen lone pair at the electron-deficient carbonyl carbon atom; protonation and elimination of water complete the reaction (Scheme 6.10).

It is not uncommon to add a trace of acid to promote these reactions.

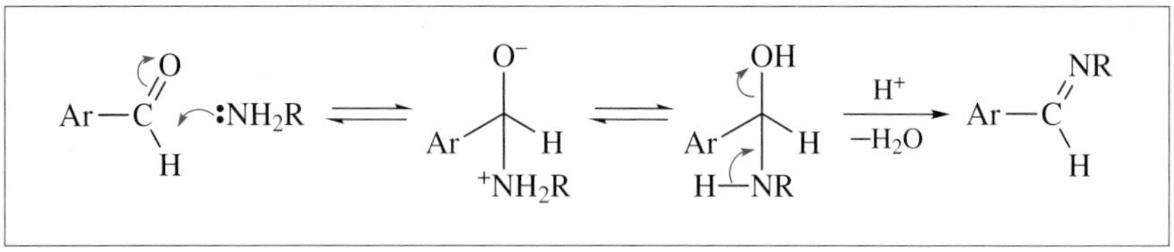

Scheme 6.10

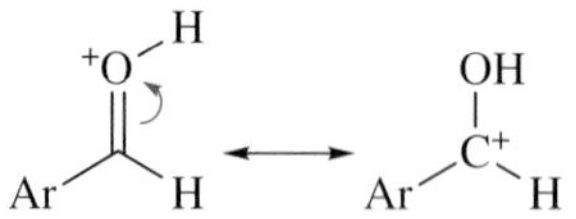

Bisulfite derivatives are used in the purification of aldehydes. A solution of the impure aldehyde is converted into its solid bisulfite compound, which is separated by filtration and so freed from impurities. Hydrolysis then liberates the purified aldehyde.

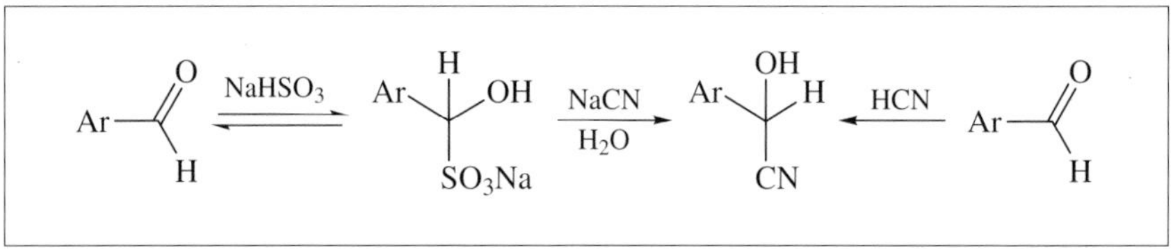

Under these conditions, the carbonyl group is protonated, effectively increasing the electron deficiency of the carbon atom.

Aromatic aldehydes react with sodium hydrogen sulfite to yield "**bisulfite**" compounds. Further reaction with sodium cyanide forms the hydroxynitrile (cyanohydrin), which can sometimes be formed directly from the aldehyde by reaction with hydrogen cyanide (Scheme 6.11).

Scheme 6.11

Thiamine (vitamin B_1) also promotes the benzoin reaction, acting initially as the nucleophile through loss of the C-2 proton and subsequently functioning as the leaving group in the final step of the reaction.

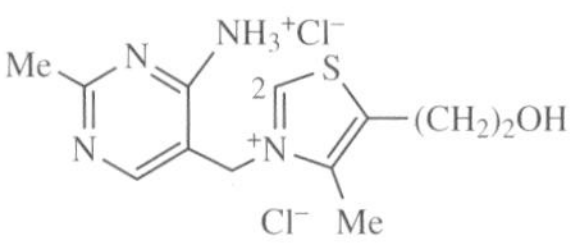

Thiamine

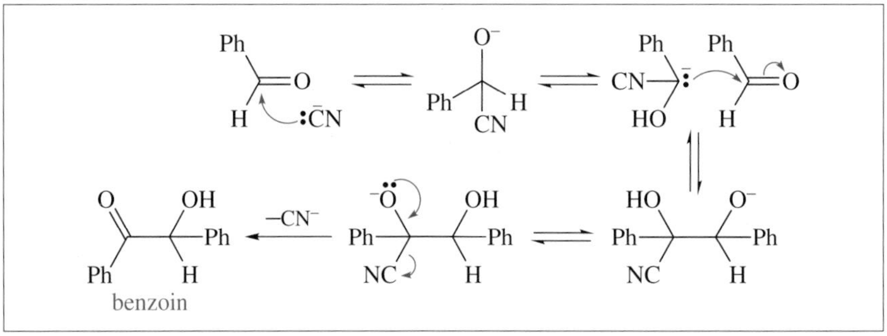

Aromatic aldehydes generally do not produce cyanohydrins on reaction with hydrogen cyanide, but undergo the **benzoin condensation** (Scheme 6.12). The initial product from nucleophilic attack by cyanide ion is deprotonated to form a resonance-stabilized carbanion, which attacks a second molecule of the aldehyde. Elimination of HCN leads to an α-hydroxy ketone, benzoin (2-hydroxy-1,2-diphenylethanone). The benzoin condensation is catalysed specifically by cyanide ion, which assists in both the formation and stabilization of the carbanion. The reaction is limited to aromatic aldehydes, since the aryl ring also stabilizes the anion.

Scheme 6.12

benzoin

In the presence of sodium hydroxide, benzaldehyde undergoes the **Cannizzaro reaction** in which two molecules of the aldehyde react to produce one molecule of benzoic acid and one molecule of benzyl alcohol (Scheme 6.13). The mechanism involves initial attack by a nucleophile, OH^- in this case, followed by hydride ion transfer.

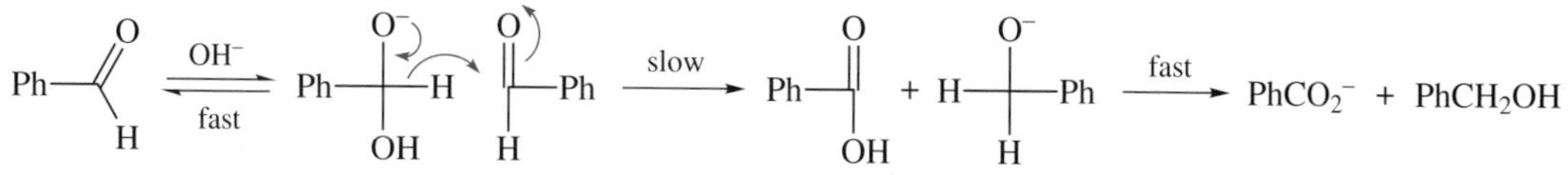

Scheme 6.13

A major structural difference between aromatic aldehydes and most aliphatic analogues is that the former lack an α-hydrogen atom. As a consequence, they are unable to enolize and so enolates/carbanions cannot be generated from them. Nevertheless, aromatic aldehydes can react with carbanions derived from, for example, aldehydes, ketones, esters and anhydrides, and so undergo a range of condensation reactions.

In the structures of benzaldehyde and phenylacetaldehyde (phenylethanal), note the lack of an α-hydrogen atom in the former compared to those indicated in the latter:

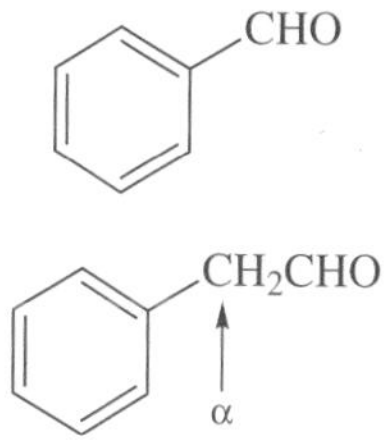

Worked Problem 6.1

Q Outline a general scheme to show how carbanions can be derived from substrates containing an α-hydrogen atom and how the carbanions react with an aromatic aldehyde.

A Mechanistically, the named reactions that follow proceed by a common pathway involving three or four steps:

1. Generation of a carbanion by abstraction of an activated α-hydrogen atom by the base. Activation results from any adjacent electron-withdrawing group. The carbanion is stabilized by resonance with the activating group, the canonical form being an enolate (Scheme 6.14).

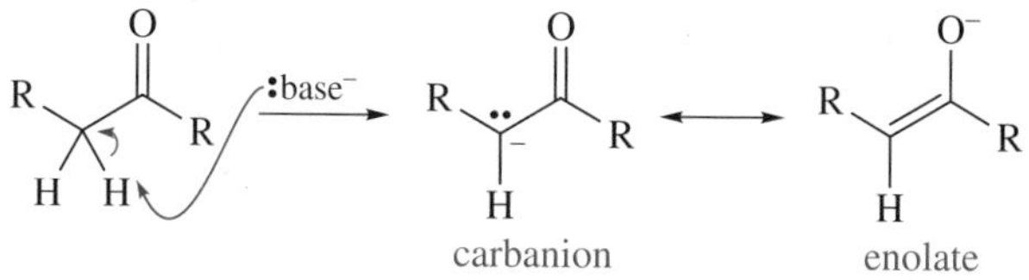

Scheme 6.14

2. Formation of a new C–C bond through attack of the carbanion at the carbonyl carbon atom of the aromatic aldehyde (Scheme 6.15).

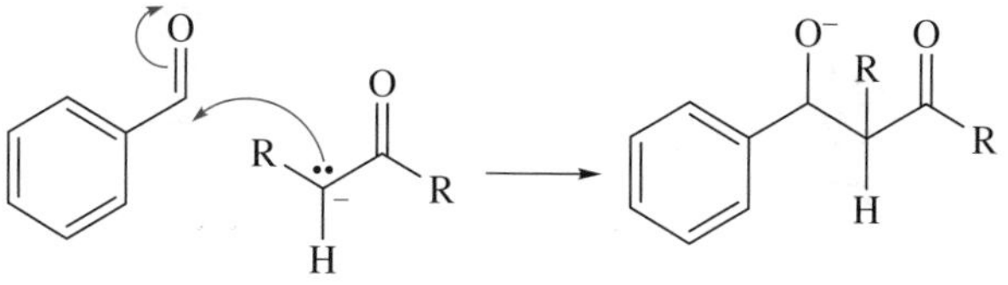

Scheme 6.15

A hydrogen atom attached to a carbon atom which is bonded to an electron-withdrawing group is said to be "activated". In a similar way, a CH_2 unit attached to a withdrawing group is called an activated methylene group. The order of activating strength of the electron-withdrawing groups is similar to that encountered in Chapter 2, when their influence on the course of electrophilic aromatic substitution was considered: NO_2 > O=CR > CN > O=COR

Aldehydes and ketones can often exist in two tautomeric forms which differ in the position of an α-hydrogen atom. The keto and enol tautomers are in dynamic equilibrium with each other. For acetone (propanone), which only contains about 10^{-7}% of the enol, the structures are:

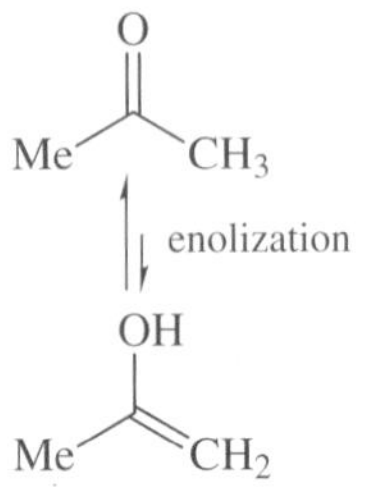

Me–C(=O)–CH_3 ⇌ (enolization) Me–C(OH)=CH_2

On the other hand, phenol exists almost entirely as the enol:

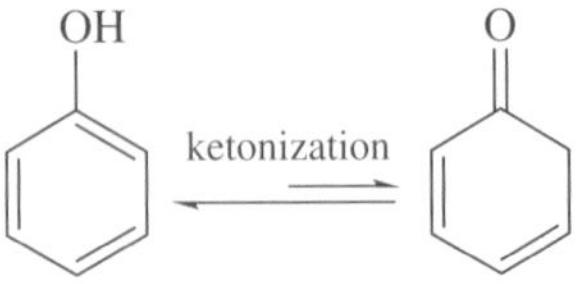

phenol (OH) ⇌ (ketonization) cyclohexadienone (O)

3. Regeneration of the base catalyst by protonation of the oxyanion formed in the previous step (Scheme 6.16).

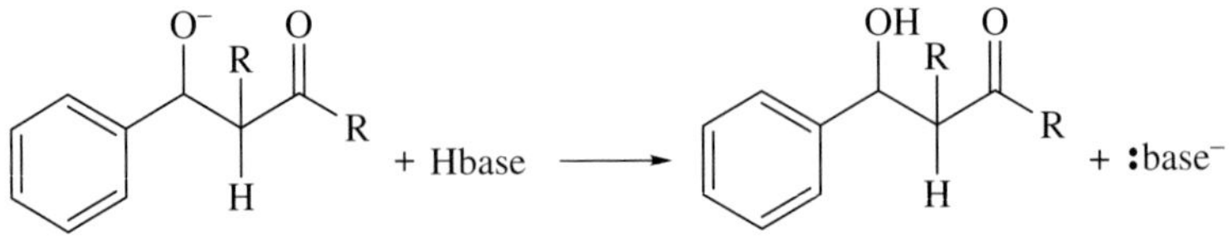

Scheme 6.16

4. Dehydration usually follows, which introduces a C=C bond (Scheme 6.17).

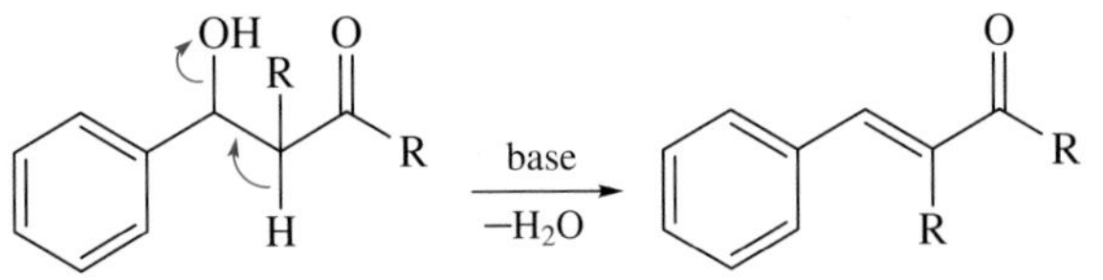

Scheme 6.17

In the **Claisen–Schmidt reaction**, aliphatic aldehydes and ketones are the sources of the carbanion and the products are unsaturated aldehydes and ketones. In the reaction with acetaldehyde (ethanal), cinnamaldehyde (3-phenylpropenal) is formed. With acetone, 4-phenylbut-3-enone (benzylideneacetone) is the product (Scheme 6.18). The reaction is generally applicable and a large variety of products have been obtained in this way, some of which are used in perfumery.

$$\mathrm{ArCH{=}CHCOMe} \xleftarrow[\mathrm{NaOH}]{\mathrm{MeCOMe}} \mathrm{ArCHO} \xrightarrow[\mathrm{NaOH}]{\mathrm{MeCHO}} \mathrm{ArCH{=}CHCHO}$$

Scheme 6.18

The synthesis of ethyl cinnamates (ethyl 3-phenylpropenoates) by the **Claisen condensation** involves generation of a carbanion, $:\bar{C}H_2CO_2Et$, from ethyl acetate or other ester (Scheme 6.19).

$$\mathrm{ArCHO} + \mathrm{MeCO_2Et} \xrightarrow{\mathrm{NaOEt}} \mathrm{ArCH{=}CHCO_2Et}$$

Scheme 6.19

The Knoevenagel reaction (Scheme 6.20) involves the reaction of aromatic aldehydes with a variety of molecules CH_2XY. The groups X and Y may be the same or different, but are invariably electron withdrawing, so creating an activated methylene group from which the carbanion $:\bar{C}HXY$ is produced. The reaction is usually carried out in pyridine solution, with piperidine as the basic catalyst. The reactions of benzaldehyde with propane-1,3-dinitrile [malononitrile, $CH_2(CN)_2$] and diethyl propane-1,3-dioate [diethyl malonate, $CH_2(CO_2Et)_2$] are illustrative. In both cases, manipulation of the $CH{=}CX_2$ group in the product allows the synthesis of other compounds.

$$PhCHO + CH_2(CN)_2 \xrightarrow{base} PhCH{=}C(CN)_2$$

Scheme 6.20

Worked Problem 6.2

Q Outline the full mechanism for the Knoevenagel reaction of benzaldehyde with diethyl malonate.

A The mechanism proceeds as outlined in Scheme 6.21, with the anion derived from diethyl malonate attacking the electron-deficient carbonyl carbon atom of benzaldehyde.

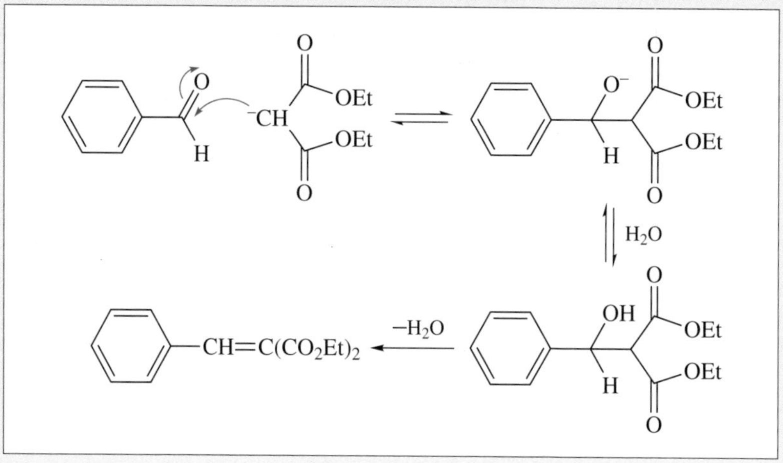

O OEt OEt O O CH H O O OEt OEt H O H₂O O OH OEt OEt H O CH=C(CO₂Et)₂ −H₂O

Scheme 6.21

In the Perkin reaction, the carbanion is generated by abstraction of an α-hydrogen from an acid anhydride, with the anion of the corresponding acid acting as the base. For example, reaction of benzaldehyde with acetic anhydride in the presence of sodium acetate at high temperature yields 3-phenylpropenoic acid (Scheme 6.22). Although the mechanism follows the general pathway, in this instance a cyclic intermediate is involved.

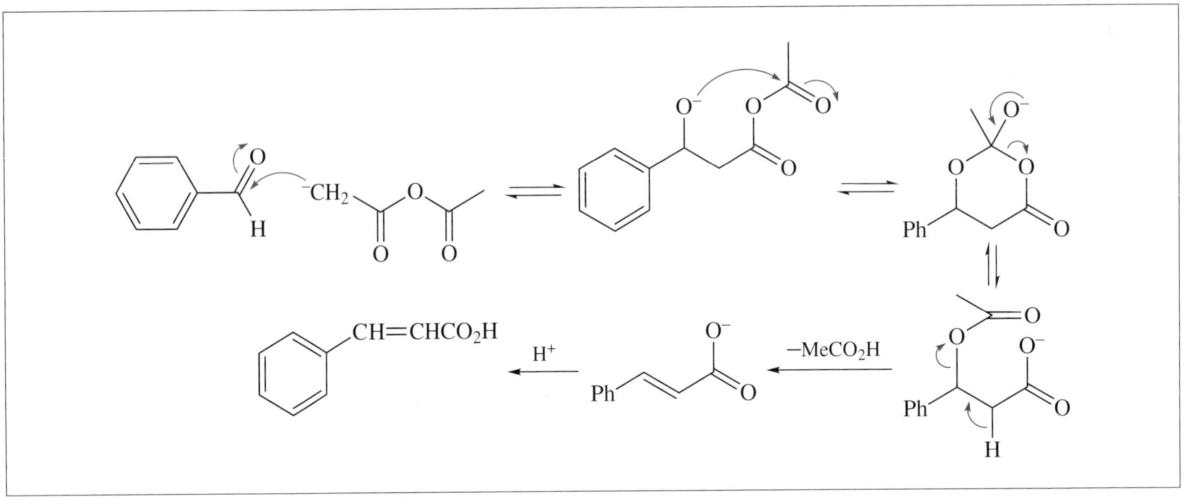

Scheme 6.22

6.4 Aromatic Ketones

6.4.1 Introduction

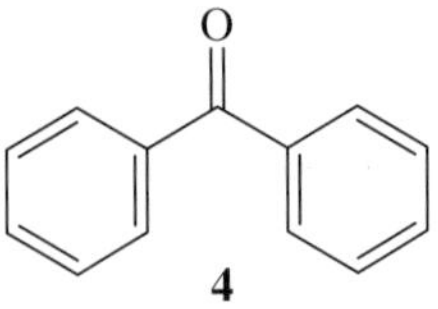

4

Aromatic ketones may contain one aryl ring and one alkyl chain, such as acetophenone (**2**), or two aryl rings such as benzophenone (diphenylmethanone, **4**). Molecules containing a carbonyl group in a side chain show normal aliphatic behaviour and are not considered here. Aromatic ketones generally behave in a similar manner to aldehydes (see Scheme 6.9) but are slightly less reactive.

6.4.2 Introduction of the Ketone Group

Both acyl and aroyl halides and anhydrides react with aromatic compounds under Friedel–Crafts conditions to yield aromatic ketones (Scheme 6.23).

$$ArH + RCOCl \xrightarrow{AlCl_3} ArCOR$$

Scheme 6.23

Highly activated aromatic compounds such as dihydric phenols can be acylated by reaction with an aliphatic nitrile in the presence of a Lewis acid, usually zinc chloride, and hydrogen chloride (Scheme 6.24). The **Houben–Hoesch reaction** is a variation of the Gattermann formylation and proceeds *via* an iminium salt, which is isolated and subsequently hydrolysed.

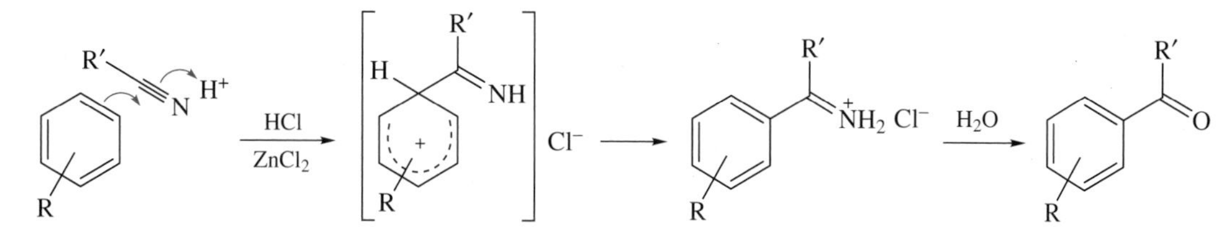

Scheme 6.24

Summary of Key Points

1. Aromatic alcohols are obtained by the hydrolysis of benzyl halides and by the reduction of aromatic aldehydes and ketones.

2. Both formyl and acyl groups can be introduced directly into an aromatic ring under Friedel–Crafts conditions to form aromatic aldehydes and ketones.

3. Both aromatic aldehydes and ketones undergo normal carbonyl group reactions with nucleophilic reagents.

4. Aromatic aldehydes have no α-hydrogen atom and so cannot form carbanions.

5. Carbanions can be generated from a variety of substrates by the base-catalysed abstraction of an α-hydrogen atom.

6. Aromatic aldehydes and ketones react with carbanions to yield aryl-substituted alkene derivatives.

Problems

6.1. Show how benzaldehyde reacts with the following, giving all steps and mechanistic details: (a) $MeNO_2$ and NaOH; (b) $PhCH_2CO_2Et$ and NaOEt; (c) PhCOMe and NaOH; (d) $PhNH_2$.

6.2. Identify the products from the following reactions, giving a mechanistic explanation for their formation: (a) 4-chlorobenzaldehyde and concentrated aqueous KOH; (b) acetophenone, formaldehyde and sodium ethoxide; (c) acetophenone and hydroxylamine; (d) benzophenone, cyclohexanone and sodium ethoxide; (e) ethyl benzoate, ethyl acetate and sodium hydride.

6.3. Suggest a route to each of the following compounds, starting from an aromatic aldehyde or ketone: (a) bis(4-methoxyphenyl)-methanol; (b) 2,3-diphenylpropenal; (c) 3-(4-nitrophenyl)propenoic acid; (d) 4-phenylbut-3-en-2-one.

7

Nitro Compounds

Aims

By the end of this chapter you should understand the following:

- How the nitro group is introduced into aromatic compounds
- Mechanistic aspects of nitration
- Reduction of the nitro group

7.1 Introduction

The nitration of benzene was discussed briefly as an example of electrophilic substitution in Chapter 2, when the **nitronium ion** (NO_2^+) was highlighted as the attacking electrophile. This reaction has probably received more mechanistic study than any other single reaction in aromatic chemistry. It is an important means of introducing functionality into the aromatic ring because the nitro group can readily be reduced to the amino group, thus providing access to many other functional groups, as described in Chapter 8.

Polynitro compounds such as 2,4,6-trinitrotoluene (TNT; 2-methyl-1,3,5-trinitrobenzene) are shock sensitive and used as explosives.

7.2 Introduction of the Nitro Group

7.2.1 Direct Nitration

The most common method of introducing a nitro group into an aromatic compound is by direct nitration and a variety of reagents have been used to achieve this. The choice of experimental conditions for the nitration of a substituted aromatic compound is based on the nature of the substituent. Thus, compounds containing an electron-withdrawing group generally require more forcing conditions and give the 3-nitro

The presence of an electron-withdrawing group in an aromatic ring makes attack by electrophiles more difficult. Electron-donor groups make this attack easier. Read Section 2.3 again if you are unsure about this.

derivative, as described in Chapter 2, whereas those substituted with an electron donor are more easily nitrated than is benzene and produce a mixture of the 2- and 4-isomers. Note that some substituents are sensitive to the oxidizing power of nitric acid.

The standard method of nitration uses a mixture of concentrated nitric acid and concentrated sulfuric acid, but when stronger conditions are required, fuming nitric acid can replace the concentrated reagent. For example, benzene is readily nitrated by mixed acid, but nitration of nitrobenzene requires fuming nitric acid in concentrated sulfuric acid (Scheme 7.1).

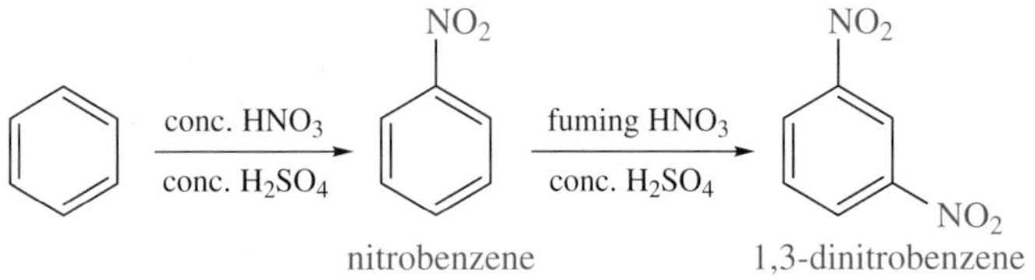

conc. HNO_3 / conc. H_2SO_4 → NO_2 nitrobenzene; fuming HNO_3 / conc. H_2SO_4 → NO_2, NO_2 1,3-dinitrobenzene

Scheme 7.1

Nitric acid alone fails to nitrate benzene and sulfuric acid also does not readily react with it, yet the mixed acid is an efficient nitrating reagent. Solutions of nitric acid in sulfuric acid show an approximately four-fold molar freezing-point depression and this has been attributed to the generation of four ions, as shown in equation (1):

Nitronium ion: $O{=}N^+{=}O$

$$HNO_3 + 2H_2SO_4 \rightleftharpoons NO_2^+ + H_3O^+ + 2HSO_4^- \qquad (1)$$

The formation of the nitronium ion is critical to the success of nitration. The purpose of the sulfuric acid is to generate the nitronium ion; other acids such as perchloric acid, hydrogen fluoride and boron trifluoride are also effective. Nitric acid alone contains only relatively small concentrations of nitronium ion and hence is not an effective nitrating agent, except with very reactive substrates.

Nitronium tetrafluoroborate is produced from NO_2F and BF_3.

The evidence for the existence of the nitronium ion is compelling. Salts such as nitronium perchlorate, $NO_2^+ClO_4^-$, and nitronium tetrafluoroborate, $NO_2^+BF_4^-$, have been isolated and successfully used as nitrating agents. A line in the Raman spectrum at 1400 cm^{-1} that originates from a linear triatomic species, which NO_2^+ is, provides spectroscopic confirmation.

When the nitration of (trifluoromethyl)benzene was carried out with nitronium tetrafluoroborate at –80 °C, Olah was able to isolate the **Wheland intermediate 1** (Scheme 7.2). Subsequent warming resulted in the formation of 1-nitro-3-(trifluoromethyl)benzene.

It is possible to nitrate highly activated aromatic compounds such as phenols using dilute nitric acid. The nitrating species is considered to be

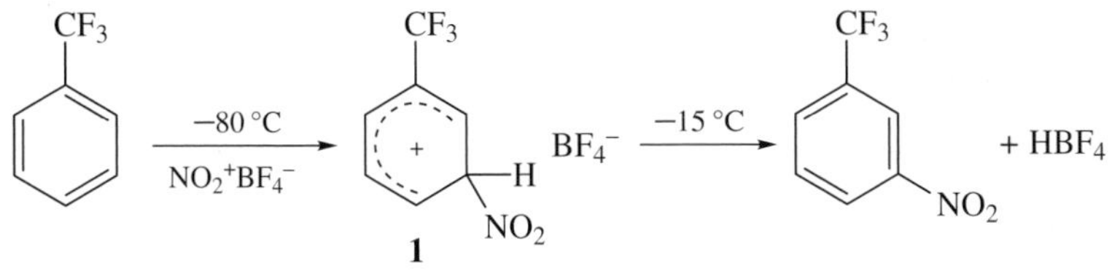

Scheme 7.2

the **nitracidium cation**, formed by interaction of two molecules of nitric acid as in equation (2):

$$2HNO_3 \rightleftharpoons H_2\overset{+}{O}{-}NO_2 + NO_3^- \qquad (2)$$

Alternatively, the reaction may proceed through nitrosation by the **nitrosonium ion**, NO^+, and subsequent oxidation of the nitroso compound by nitric acid (Scheme 7.3). There is only a small concentration of NO^+ in dilute nitric acid and so catalytic amounts of sodium nitrite are sometimes added to increase the quantity. This technique is a useful means of effecting smooth, low-temperature nitrations.

Scheme 7.3

$$HNO_2 + 2HNO_3 \rightleftharpoons H_3O^+ + 2NO_3^- + NO^+$$

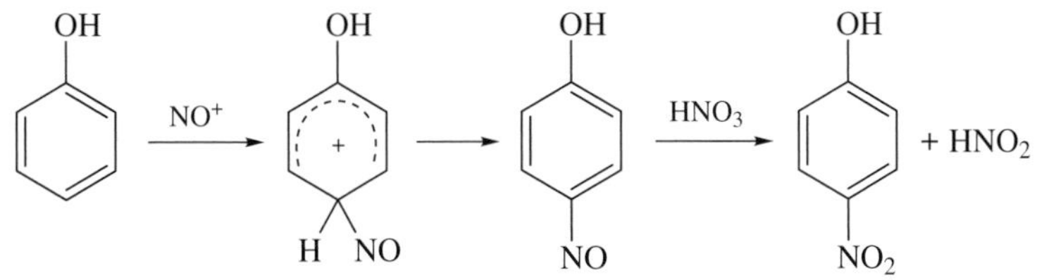

Solutions of nitric acid in acetic acid or acetic anhydride are effective nitrating agents, generating acetyl nitrate. The exact nature of the nitrating agent has not been confirmed, but it is suggested that, of the possible species involved, N_2O_5 and the nitronium ion are the most likely. A high proportion of 2-substitution occurs when this reagent is used with substrates containing a functional group bonded to the ring through a heteroatom. An example is the nitration of methoxybenzene, in which 70% of the 2-nitro isomer is formed. It is proposed, in this instance, that the nitrating species is dinitrogen pentoxide and that the mechanism involves a six-centre cyclic rearrangement (Scheme 7.4).

Safety note: nitration reactions generate significant quantities of heat (~200 kJ mol^{-1}). Temperature control is therefore essential in order to prevent polynitration and to avoid self-generating runaway reactions that can have catastrophic consequences. With due care, nitrations are perfectly safe and are commonly carried out on a very large scale in industry. Indeed, mild nitrations carried out in 80% acetic acid, for example, are very safe. However, literature methods should be carefully followed. New nitrations should be carried out initially on a very small scale and with extreme care. Nitrations are rarely carried out above 120–140 °C; reactions at higher temperatures would be considered hazardous.

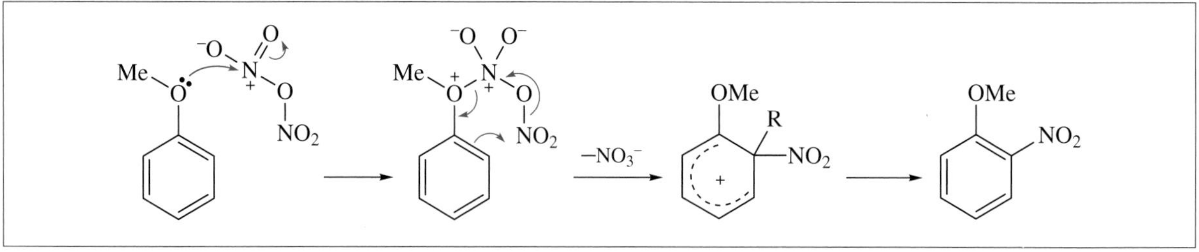

Scheme 7.4

Worked Problem 7.1

Q What products result when (a) ethylbenzene, (b) methyl benzoate and (c) *N,N*-dimethylaniline are nitrated with mixed acid?

A The electronic properties of the substituent attached to the benzene ring usually determine the site of attack by the nitronium ion. (a) An ethyl group is weakly electron donating and so is an *ortho/para* director. On nitration, ethylbenzene gives approximately equal amounts of 1-ethyl-2-nitrobenzene and 1-ethyl-4-nitrobenzene. (b) An ester group, CO_2R, is electron withdrawing and so directs electrophilic attack to the *meta* position. Hence, methyl benzoate is converted into methyl 3-nitrobenzoate. (c) Although an amino group is strongly electron donating, under the acidic conditions of nitration it is initially protonated at nitrogen. Here, the dimethylamino group is converted into the dimethylammonium group, $Me_2\overset{+}{N}H–$. The lone pair of electrons has been used to form the bond to the proton and so is no longer available for donation to the aromatic π-cloud. The increased electronegativity of the positively charged nitrogen atom makes the group strongly electron withdrawing and *meta* directing. The product is therefore *N,N*-dimethyl-3-nitroaniline after basification of the reaction mixture.

7.2.2 Indirect Methods of Introducing the Nitro Group

The conversion of an amino group into a nitro group can be useful when specific substitution patterns are required. The synthesis of 1,4-dinitrobenzene from 4-nitroaniline is illustrative (Scheme 7.5). Oxidation can be accomplished directly using peroxytrifluoroacetic acid or in two steps using H_2SO_5 (monoperoxysulfuric acid) and oxidation of the resulting nitroso compound with hydrogen peroxide. Alternatively, the amine can be diazotized in fluoroboric acid and then reacted with sodium nitrite in the presence of copper powder.

H_2SO_5 is formed when 30% hydrogen peroxide solution is added to concentrated sulfuric acid. It is known as Caro's acid.

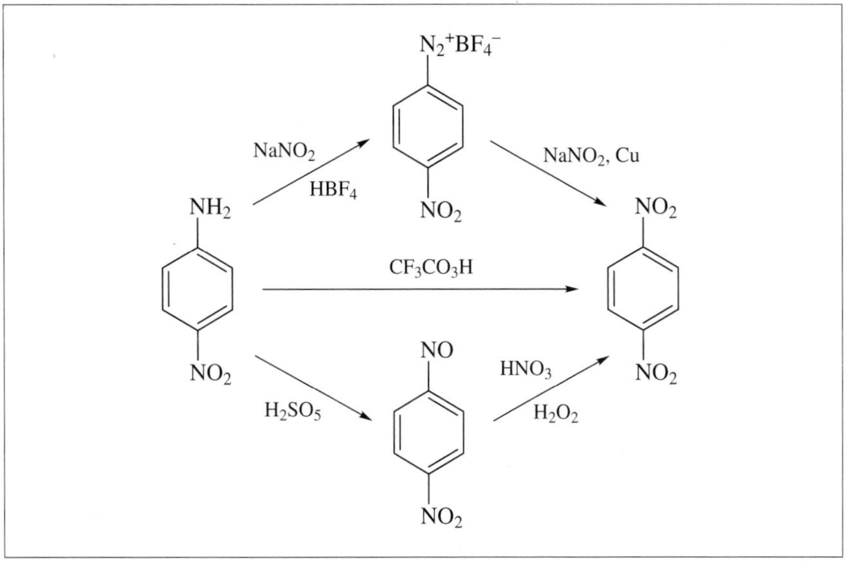

Scheme 7.5

7.3 Charge Transfer Complexes

Polynitro compounds form charge transfer or π-complexes with certain aromatic hydrocarbons. The presence of electron-withdrawing groups in the ring of the electron acceptor and electron-donating groups in the ring of the electron donor are essential for the formation of these complexes. Such a complex is formed between 1,3,5-trimethylbenzene (mesitylene) and 1,3,5-trinitrobenzene. In mesitylene, the three methyl groups exert a +I effect to enhance the electron density of the π-electron cloud, but in 1,3,5-trinitrobenzene the opposite effect occurs and the π-electron cloud is relatively electron deficient. The result is the formation of the charge transfer complex **2**, in which the two rings lie in approximately parallel planes, and a weakly bonded, coloured compound results.

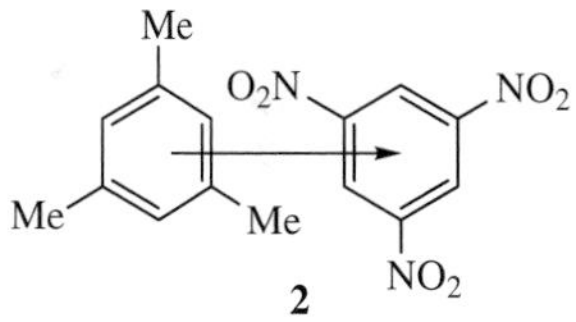

7.4 Reactions of Nitro Compounds

7.4.1 Reactions of the Ring

The nitro group is strongly electron withdrawing and as such is *meta* directing and deactivating towards further electrophilic attack. Conversely, this electron withdrawal activates the ring to nucleophilic attack and progressive introduction of nitro groups into aryl halides makes the displacement of halogen by nucleophiles easier (see Chapter 9).

7.4.2 Reactions of the Nitro Group

Reduction of a nitro group to an amino group is one of the most important reactions in aromatic chemistry and there are many methods available to achieve this transformation.

The method of choice in industry is **catalytic hydrogenation** under pressure. The nitro compound, dissolved in an alcohol such as propan-2-ol, and in the presence of a finely divided metal catalyst, is maintained in an atmosphere of hydrogen gas, usually under pressure in an autoclave. On completion of the reaction, the vessel is vented and the catalyst is filtered off, leaving simply the product in the solvent. The catalysts used are usually either finely divided platinum or palladium suspended on carbon or alumina, or the Raney catalysts which are finely divided nickel or, more recently, copper and cobalt. The efficiency of these catalysts can be seriously reduced by poisoning and care has to be taken to ensure that sulfur compounds and other poisons are excluded from the reaction mixture. Raney nickel is less susceptible to sulfur poisoning. The choice of catalyst and reaction conditions are important. For example, the reduction of halonitro compounds with Raney nickel can result in significant dehalogenation as well as reduction of the nitro group, but the use of platinum on carbon effectively eliminates this problem.

The metal catalysts are recovered and regenerated by the manufacturers. Care is required when handling finely divided metal catalysts since some of them react readily with atmospheric oxygen and burst into flames.

Commercial use of this method is increasing and some companies have developed this technique to a high degree of sophistication.

Catalytic hydrogenation is a heterogeneous reaction which occurs at the surface of the catalyst. The mechanism is complex and proceeds through the nitroso and hydroxylamine derivatives; minor by-products such as the azo and azoxy compounds occasionally appear.

In the absence of a laboratory or industrial pressure vessel, **catalytic hydrogen** transfer is a viable alternative. Here, the nitro compound is stirred in a solvent with a catalyst, commonly 1% platinum on carbon, in the presence of a hydrogen donor such as hydrazine, sodium formate (sodium methanoate) or methylcyclohexene (Scheme 7.6). In the case of hydrazine, interaction at the catalyst surface produces the reductant, hydrogen, and nitrogen. Toluene is the by-product when methylcyclohexene is used.

$$\text{PhNO}_2 + \text{1-methylcyclohexene} \xrightarrow{1\%\ \text{Pt/C}} \text{PhNH}_2 + \text{PhMe}$$

Scheme 7.6

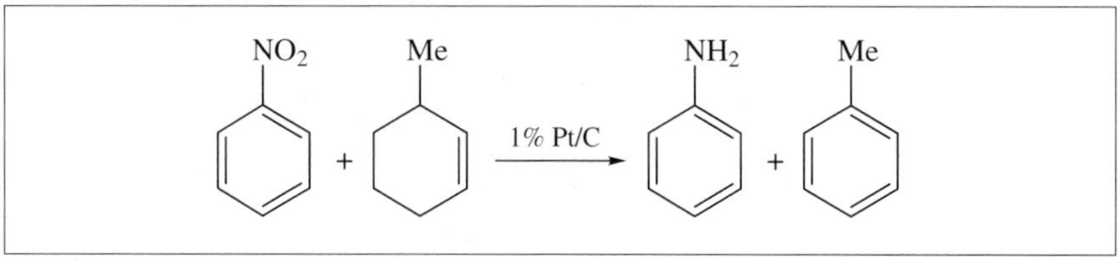

Various combinations of **metal and acid** reduce nitro groups and this method lends itself to both small-scale laboratory work and industrial conditions. Tin and hydrochloric acid are used when small quantities are involved, but work-up requires basification to destroy the amine–chlorostannate complex. Tin is also an expensive metal. The use of zinc and hydrochloric acid also requires a basic work-up. In both cases the disposal and handling of large amounts of residues present an environmental problem which industry seeks to avoid. Reduction by iron and HCl or acetic acid is suitable for most applications and offers the advantage that often only a catalytic amount of acid is required. The iron(III) oxide by-product precipitates and so is easy to remove.

Cost, convenience and environmental considerations have to be considered by industry in addition to the usual factors of yield and compatibility of the substrate with the reaction conditions.

Other reducing agents such as sodium hydrosulfite (sodium dithionite, $Na_2S_2O_4$) and sodium sulfide have also been used. For example, the latter reduces one of the nitro groups in polynitro compounds selectively; 1,3-dinitrobenzene gives 3-nitroaniline in this way (Scheme 7.7). Tin(II) chloride reduces nitro groups selectively in the presence of carbonyl groups under acidic conditions.

Titanium(III) chloride also reduces the nitro group to the amino group in acid solution. This reaction is the basis of a quantitative method for the determination of nitro groups.

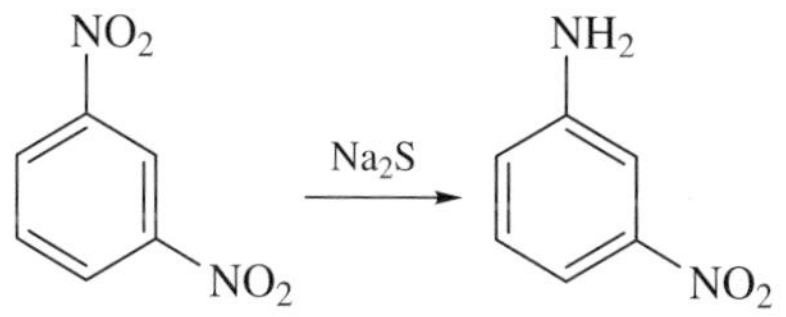

NO_2 / NO_2 $\xrightarrow{Na_2S}$ NH_2 / NO_2

Scheme 7.7

7.5 Nitrosobenzene and Phenylhydroxylamine

Nitroso compounds are of relatively limited importance in aromatic chemistry. However, since the nitroso group is easily reduced to an amino group, they do offer an indirect route to aromatic amines. The nitrosation of phenols and dialkylanilines and the subsequent reduction are pertinent examples. The nitroso group is also readily oxidized to a nitro group.

Nitrosation of dialkylanilines followed by reduction produces *N,N*-dialkyl-*p*-phenylenediamines, which are used as developers in colour photography.

Worked Problem 7.2

Q The nitroso group exhibits similar behaviour to a carbonyl group. Predict the products from the reaction of nitrosobenzene with (a) malononitrile (propane-1,3-dinitrile), (b) hydroxylamine and (c) aniline. Suggest a mechanism for the last reaction.

A The reactions proceed in exactly the same way as with the carbonyl group, by loss of water and generation of a double bond (Scheme 7.8). The mechanism for the reaction with an amine is illustrated in Scheme 7.9.

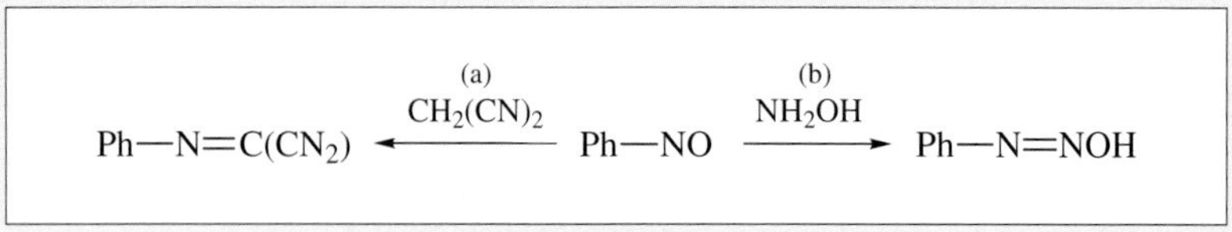

(a) $CH_2(CN)_2$ (b) NH_2OH

Ph—N=C(CN$_2$) ← Ph—NO → Ph—N=NOH

Scheme 7.8

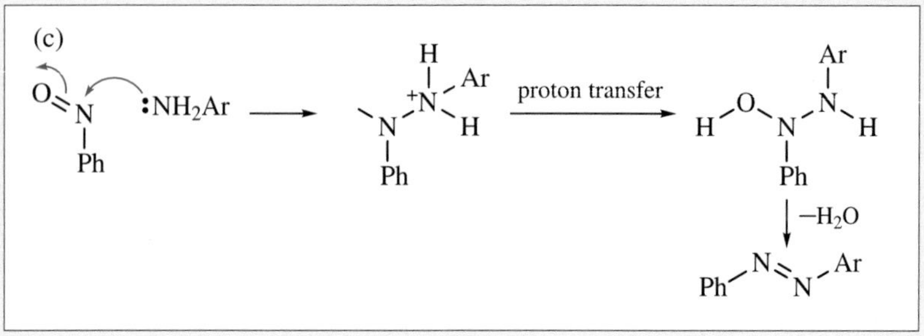

(c)

O=N–Ph + :NH$_2$Ar → (Ph)N(O⁻)–$^+$N(H)(H)Ar —proton transfer→ H–O–N(Ph)–N(H)Ar —(−H$_2$O)→ Ph–N=N–Ar

Scheme 7.9

Nitrosobenzene (**3**) can be prepared by oxidation of aniline by H_2SO_5 (see Section 7.2.2) and by oxidation of phenylhydroxylamine (**4**) with potassium dichromate. Phenylhydroxylamine is available from nitrobenzene by reduction with zinc dust and aqueous ammonium chloride (Scheme 7.10).

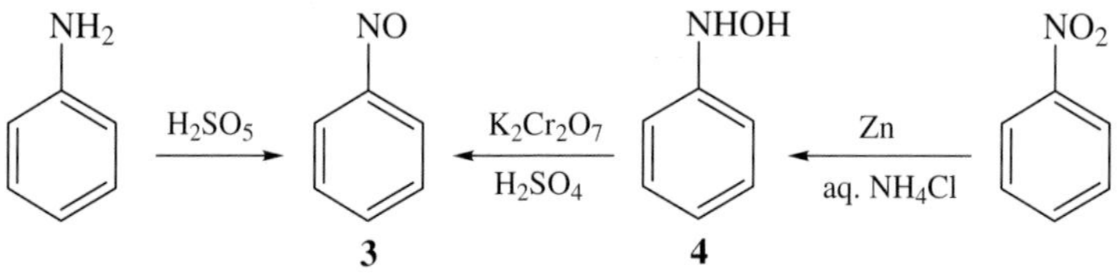

NH_2 —H_2SO_5→ NO (**3**) ←$K_2Cr_2O_7$, H_2SO_4— NHOH (**4**) ←Zn, aq. NH_4Cl— NO_2

Scheme 7.10

Summary of Key Points

1. Nitration is brought about by the nitronium ion, NO_2^+.

2. The nitro group is strongly electron withdrawing and directs electrophilic attack, which is significantly more difficult than for benzene, to the *meta* position.

3. Reduction of a nitro group to an amine can be achieved by catalytic hydrogenation, hydrogen transfer and by various combinations of metal and acid.

4. Partial reduction of a nitro group to a hydroxylamino function occurs on treatment with zinc dust and aqueous NH_4Cl.

5. Activated aromatic compounds can be nitrosated by the nitrosonium ion, NO^+.

6. Nitroso aromatic compounds can be derived by oxidation of amines and hydroxylamines.

Problems

7.1. Arrange the following compounds in order of their reactivity towards nitration and give the formulae of the major products of the reactions: (a) benzene, benzonitrile, methoxybenzene, (trifluoromethyl)benzene; (b) acetanilide, benzoic acid, chlorobenzene, acetophenone; (c) phenol, ethyl benzoate, nitrobenzene, bromobenzene.

7.2. Suggest synthetic routes to the following compounds, starting from the indicated starting material: (a) 1-(3-nitrophenyl)propan-1-one from benzene; (b) 1,4-dinitrobenzene from benzene; (c) 2-nitroaniline from aniline.

8 Aromatic Amines and Diazonium Salts

Aims

By the end of this chapter you should understand:

- How an amino group can be introduced into aromatic molecules
- The reactions of the amino group
- The basicity of amines
- The reactions of diazonium salts

NH_2 CH_2NH_2 NHMe

1 2 3

NMe_2 NH_2

4 Me 5

Trivial names for simple *C*-alkylated anilines are in common use; thus the monomethylanilines (monomethylphenylamines) are known as toluidines (*e.g.* ***p*-toluidine**, **5**), and the dimethylanilines (dimethylphenylamines) are called xylidines.

8.1 Introduction

Aromatic amines fall into two categories: those in which the amino group is directly attached to the aromatic ring and those where an amino group is part of a side chain. This chapter concentrates on the former group and deals with compounds such as **aniline** (phenylamine, benzenamine **1**), rather than benzylamine (phenylmethanamine, **2**) and similar compounds in which the amino group exhibits the reactions of an aliphatic amine.

Aromatic amines are subdivided into **primary amines**, $ArNH_2$, such as aniline, **secondary amines**, ArNHR, in which the amino group has been monoalkylated and typified by *N*-methylaniline (*N*-methyl-*N*-phenylamine, **3**), and **tertiary amines**, $ArNR_2$, such as *N*,*N*-dimethylaniline (*N*,*N*-dimethyl-*N*-phenylamine, **4**).

8.2 Introduction of the Amino Group

8.2.1 Reductive Methods

In Chapter 7, the reduction of the nitro group to produce aromatic amines was discussed in some detail. This is the most important method for synthesizing amines.

Other groups such as nitroso, hydroxylamine, azo, azoxy and hydrazo may also be reduced by similar means, though with varying degrees of difficulty, but their use is rather limited.

ArNO	nitroso
ArNHOH	hydroxylamine
ArN=NAr	azo (diazeno)
$ArN^+(O^-)=NAr$	azoxy (diazene oxide)
ArNHNHAr	hydrazo (diazano)

8.2.2 Molecular Rearrangements

There is a series of related reactions in which the common theme is a 1,2-shift of an aryl fragment from C to an electron-deficient nitrogen atom. These named reactions provide a useful means of introducing an amino group. The general mechanistic picture for these reactions is shown in Scheme 8.1. An actual intermediate, a nitrene, in which a nitrogen atom has only a sextet of electrons, is not proposed, but rather the shift of the arene portion R to the nitrogen atom is concerted with the departure of an electron-rich leaving group X from the nitrogen.

A nitrene is a species in which the nitrogen has only a sextet of electrons (cf. carbene, p.54): R–N̈:

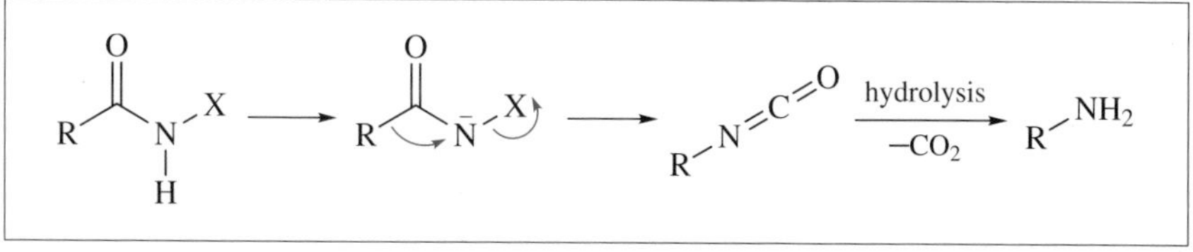

$$RC(=O)NHX \longrightarrow RC(=O)N^{-}X \longrightarrow R{-}N{=}C{=}O \xrightarrow[-CO_2]{\text{hydrolysis}} R{-}NH_2$$

Scheme 8.1

The individual named reactions are summarized in Table 8.1, but two examples are particularly useful and are discussed in detail.

Table 8.1 Synthesis of aromatic amines by rearrangements involving migration from C to an electron-deficient N atom

Reaction name	*Conditions*	*Leaving group (X)*
Hofmann	$ArCONH_2 + Br_2/NaOH \rightarrow ArNH_2$	Br^-
Curtius	$ArCONHNH_2 + HONO \rightarrow ArNHCOMe$	N_2
Lossen	$ArCONHOH + HCl \rightarrow ArNH_2$	H_2O
Schmidt	$ArCO_2H + HN_3 \rightarrow ArNH_2$	N_2
Beckmann	$Ar(Me)C{=}NOH + H^+ \rightarrow ArNHCOMe$	H_2O

In the **Hofmann reaction** (Scheme 8.2), an amide is treated with a halogen, usually bromine, in alkali in which the reactive species is the hypohalite ion (XO^-). This reaction, which applies to aliphatic as well as aromatic amides, has been thoroughly studied. Evidence supporting

Much of the mechanistic study involved aliphatic amides and evidence for the concerted mechanism includes the fact that there is complete retention of stereochemistry in the migrating aliphatic group.

the mechanism includes the isolation and identification of the intermediate *N*-bromoamide and the isocyanate. A key feature of the mechanism is the role of the halogen. It not only increases the acidity of the amide proton, but also acts as a good leaving group, prompting the rearrangement. The isocyanate reacts with water and the resulting unstable carbamic acid spontaneously decarboxylates to the amine.

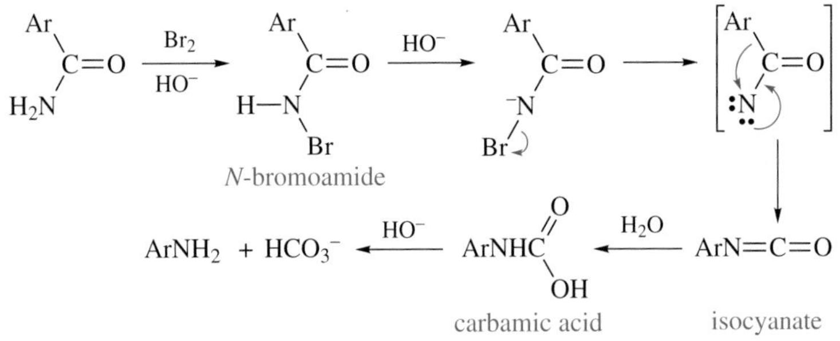

N-bromoamide

carbamic acid

isocyanate

Scheme 8.2

Probably the best-known rearrangement in which an aryl group migrates from carbon to nitrogen is the **Beckmann rearrangement**. Here, ketoximes are converted to *N*-substituted amides when treated with an acidic reagent, such as sulfuric or polyphosphoric acid, phosphorus pentachloride or thionyl chloride ($SOCl_2$). In the context of this chapter, the amides are hydrolysed to liberate the amine. The mechanism in acid media is believed to proceed as illustrated in Scheme 8.3, where the 1,2-shift from carbon to nitrogen is noted as the key step.

Oximes can exist as geometrical isomers, although only one isomer is usually formed during their synthesis. The Beckmann rearrangement is stereospecific and only the group (Ar) that is *anti* to the leaving group, the hydroxy function, migrates. Thus, acetophenone oxime (Ar = Ph) rearranges only to acetanilide.

It is possible to determine the stereochemistry of an oxime by identifying the Beckmann product. For example, hydrolysis of the product from the Beckmann rearrangement of the oxime derived from 4-chloroacetophenone gives 4-chloroaniline and not 4-chlorobenzoic acid. This is because the hydroxyl group of the oxime is *anti* to the aryl ring.

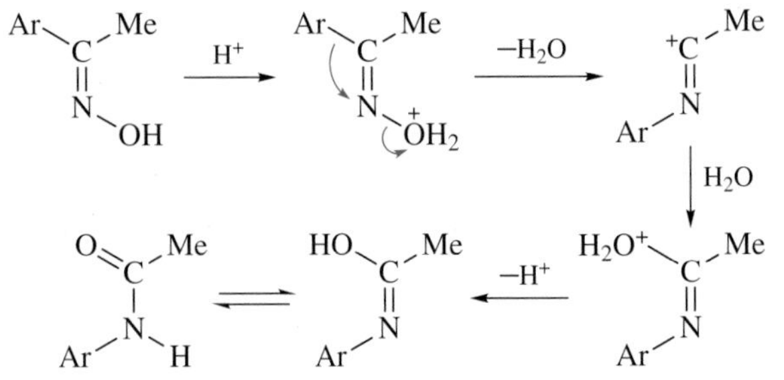

Scheme 8.3

Worked Problem 8.1

Q Suggest a route to aromatic amines from benzene that involves acylation.

A Acetophenone derivatives can be prepared from benzene by a Friedel–Crafts acylation. The derived oxime, formed using hydroxylamine, undergoes a Beckmann rearrangement to produce the amide; hydrolysis yields the amine (Scheme 8.4).

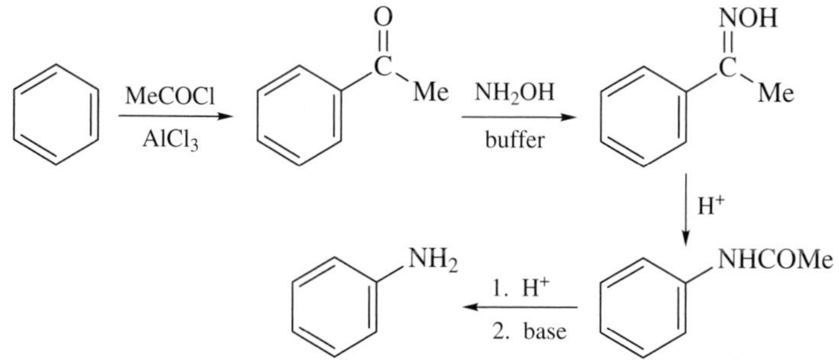

Scheme 8.4

8.2.3 Miscellaneous Methods

Aryl halides in which electron-withdrawing groups activate the halogen atom to nucleophilic displacement react directly with ammonia to produce amines. Non-activated aryl halides yield amines on reaction with sodamide through the intermediacy of an aryne intermediate (see Chapter 9).

8.3 Reactions of Aromatic Amines

8.3.1 Reactions of the Ring

The amino group is strongly electron donating and directs substitution to the *ortho* and *para* positions. This is illustrated by the addition of chlorine or bromine water to aniline, which results in immediate reaction and the precipitation of the 2,4,6-trihalogenated derivative.

The amino group is protonated when reactions are carried out in acidic media. Formation of the anilinium salt, $ArNH_3^+$, significantly alters the influence of the group on ring reactivity. The protonated group is strongly electron withdrawing (–I effect) and is *meta* directing. However, even in strongly acidic conditions there exists an equilibrium between the anilinium cation and the free base. Consequently, a fast reaction with the *ortho/para* directing free base competes with a slow *meta* directing reaction of the more abundant anilinium cation and mixtures result. Reactions such as sulfonation and nitration therefore proceed in a more controlled fashion when carried out on the acylated amine.

8.3.2 Reactions of the Amino Group

The amino group is a reactive species and undergoes some important reactions.

Acetanilide can be prepared by reaction of aniline with acetic anhydride or acetyl chloride (Scheme 8.5). This reaction occurs with most anhydrides and acyl chlorides to produce anilides.

The 4-isomer is the major product because attack at the 2-position is sterically hindered. Nevertheless, *ca.* 10% of 2-bromoacetanilide is produced during the bromination of acetanilide.

The electron-donating ability of an amino group is moderated by acetylation. Only monobromination of acetanilide occurs on reaction with bromine in acetic acid (Scheme 8.5). Acetanilide behaves similarly on nitration.

$PhNH_2 \xrightarrow{Ac_2O} PhNHCOMe \xrightarrow[80\%]{Br_2,\ AcOH} 4\text{-}BrC_6H_4NHCOMe$

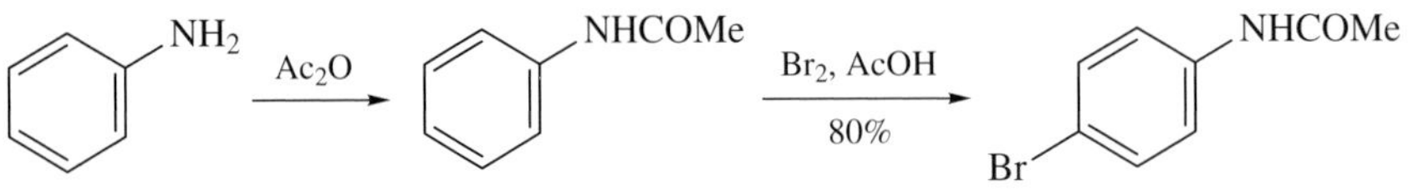

Scheme 8.5

$H(Ph)\ddot{N}{-}C(=O)Me \leftrightarrow H(Ph)\overset{+}{N}{=}C(O^-)Me$

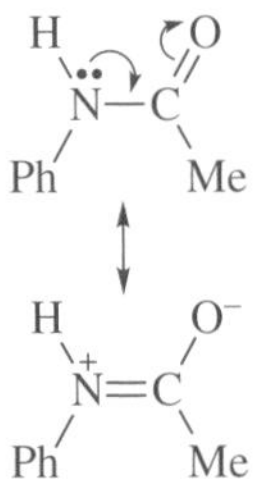

Acylation is a means of controlling isomer formation and of moderating the electron-donating capacity of the group and so preventing polysubstitution. The carbonyl function attracts the nitrogen lone pair of electrons, which is therefore less available for resonance interaction with the π-system of the ring. This also accounts for the lack of carbonyl group reactions in amides.

The amino group can be *N*-alkylated with iodomethane (Scheme 8.6) to give initially *N*-methylaniline and then *N*,*N*-dimethylaniline. The final product is the trimethylammonium salt, which is formed by quaternization of the nitrogen.

A quaternary atom has four organic substituents attached to it. Quaternization is the process of attaching the fourth substituent.

$ArNH_2 \xrightarrow{MeI} ArNHMe \xrightarrow{MeI} ArNMe_2 \xrightarrow{MeI} Ar\overset{+}{N}Me_3\ I^-$

Scheme 8.6

Amines can be reductively alkylated by aldehydes and ketones in the presence of hydrogen and a hydrogenation catalyst.

N-Methylaniline may be formylated with formic acid (methanoic acid, HCO_2H). The product, *N*-methylformanilide (**6**), is used as a formylating reagent.

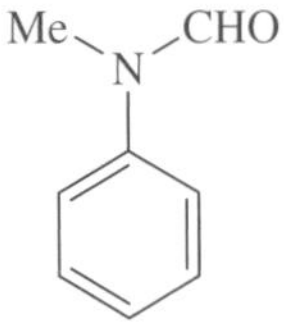

$PhN(Me)CHO$

6

Worked Problem 8.2

Q Explain how an *N*-isopropyl group might be introduced into aniline by a reductive alkylation technique.

A Reaction of aniline with acetone in the presence of catalyst and hydrogen gas results in the production of *N*-isopropylaniline *via* reduction of the initially formed imine (Scheme 8.7).

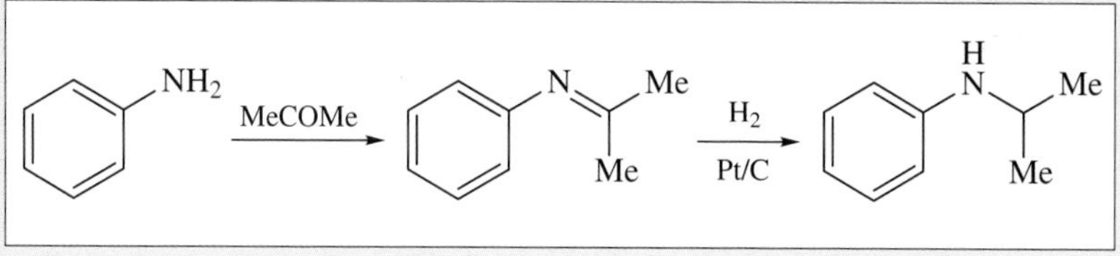

Scheme 8.7

The reaction of aromatic amines with nitrous acid is of considerable importance and the formation of diazonium salts from the primary amines is discussed in detail in Section 8.6. Reaction of nitrous acid with secondary amines does not give diazonium salts, but results instead in *N*-nitrosation. Tertiary amines such as *N*,*N*-dimethylaniline do not *N*-nitrosate, but undergo electrophilic substitution by the nitrosonium cation (NO^+) to give *N*,*N*-dimethyl-4-nitrosoaniline (Scheme 8.8).

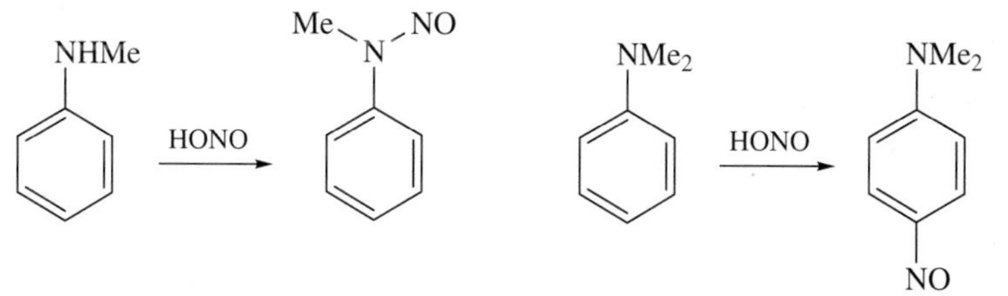

Scheme 8.8

8.4 Related Compounds

There are several compounds which require different methods of synthesis or which show special reactions.

Diphenylamine (**7**) is prepared industrially either by heating aniline with aniline hydrochloride at 140 °C under pressure, or by heating aniline with phenol at 260 °C in the presence of zinc chloride. The most convenient laboratory synthesis uses the **Ullmann reaction** (Scheme 8.9) (see Chapter 10), in which acetanilide is refluxed with bromobenzene in the presence of potassium carbonate and copper powder in nitrobenzene solvent. Triphenylamine is similarly prepared from diphenylamine and iodobenzene.

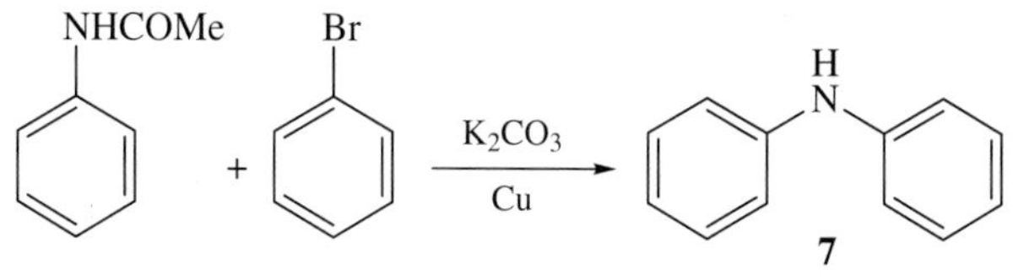

Scheme 8.9

Benzylamine (**2**), which behaves as an aliphatic amine, may be prepared by the reduction of benzonitrile or benzaldoxime (Scheme 8.10).

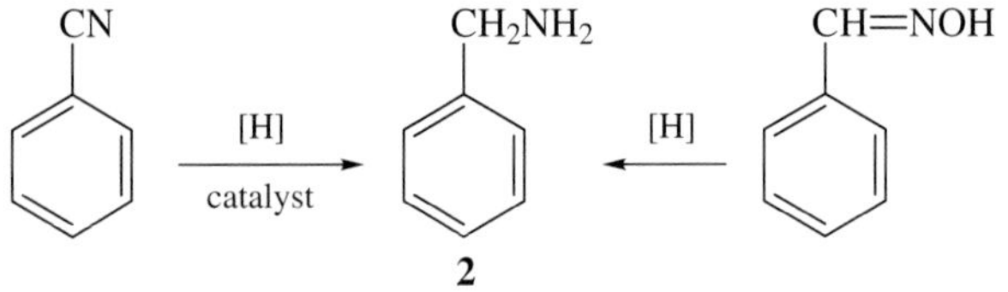

CN → [H] / catalyst → CH_2NH_2 (**2**) ← [H] ← CH=NOH

Scheme 8.10

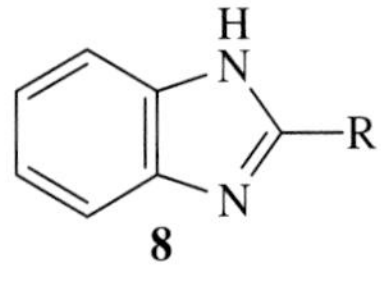

8 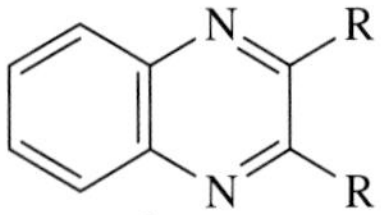 **9** 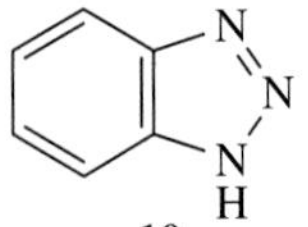 **10**

The diaminobenzenes (phenylenediamines) are prepared by reduction of 1,3-dinitrobenzene and 2- and 4-nitroanilines. *o*-Phenylenediamine is of value in the synthesis of a range of nitrogen heterocycles. Thus reaction with organic acids produces benzimidazoles (**8**). With 1,2-dicarbonyl compounds, quinoxalines (**9**) are produced. Treatment with nitrous acid results in diazotization of one amino group followed by immediate cyclization to give benzotriazole (**10**).

8.5 Basicity of Amines

Aniline is a weaker base than aliphatic amines because the lone pair of electrons is conjugated with the π-system of the aromatic ring and structures such as **11–13** contribute to the actual structure of aniline. Thus, the lone pair of electrons on the nitrogen atom is less available for coordination to a proton. The effect is considerable and aniline has a pK_a of 4.6 compared to methylamine's pK_a of 10.6; it thus reduces the basicity a million-fold.

Aniline forms stable, crystalline salts with strong mineral acids. This is a convenient way of storing amines which otherwise tend to oxidize and darken on exposure to air.

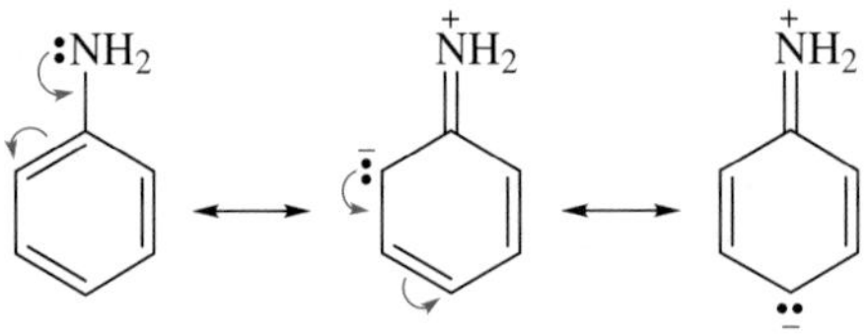

$:NH_2$ ⟷ $\overset{+}{N}H_2$ ⟷ $\overset{+}{N}H_2$

11 **12** **13**

Electron-donating substituents increase the basicity of aromatic amines, although the effect is not very great as can be seen from the data presented in Table 8.2 for the methyl group. Electron-withdrawing groups have a more pronounced weakening influence on basicity. In both cases, the position of the substituent and its ability to enter into resonance with the amino group also has an effect on the base strength. For example, the nitroanilines are all weaker bases than aniline itself because of the strong mesomeric (–M) and inductive (–I) electron-withdrawing effects of the nitro group. In the 2- and 4-positions, both effects are operative. Canonical forms such as **14** and **15** contribute to the structure and illustrate the involvement of the lone pair of electrons on the nitrogen atom with the π-system of the aromatic ring and with the nitro group.

The –M effect cannot operate in the case of 3-nitroaniline, but this is still a weaker base than aniline because of the –I effect, although not as weak as the 2- and 4-isomers (Table 8.2). A second nitro group enhances these effects and 2,4-dinitroaniline is so weak a base that it does not dissolve in dilute hydrochloric acid.

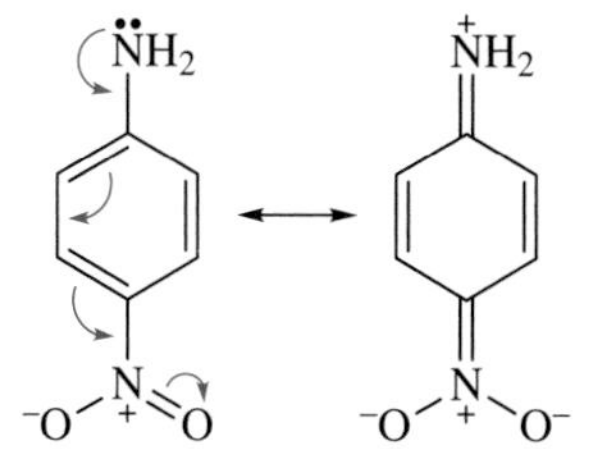

NH_2 NH_2 14 15

The trivial name for 2,4,6-trinitroaniline is picramide, which tells us that it lacks basic character. Its properties are more like those of an amide than those of an amine.

Table 8.2 pK_a values of some substituted anilines

Substituent	*2-*	*3-*	*4-*
Me	4.39	4.69	5.12
NO_2	–0.29	2.50	1.02
MeO	4.49	4.20	5.29
Cl	2.64	3.34	3.98

The methoxy group is somewhat unusual in that in the 4-position its +M effect increases basicity, but only a –I effect operates in the 3-position and this decreases the basicity.

Substituents in the *ortho* position may exert a steric effect on the amino group, twisting it out of the plane of the ring and so reducing mesomeric interaction between the nitrogen lone pair and the ring. The lone pair is more available for donation to a proton and the basicity is therefore greater.

8.6 Diazonium Salts

Treatment of a primary aromatic amine, such as aniline, dissolved or suspended in an aqueous mineral acid, with aqueous sodium nitrite solution whilst the temperature is maintained below 5 °C, produces the relatively unstable diazonium salt (Scheme 8.11).

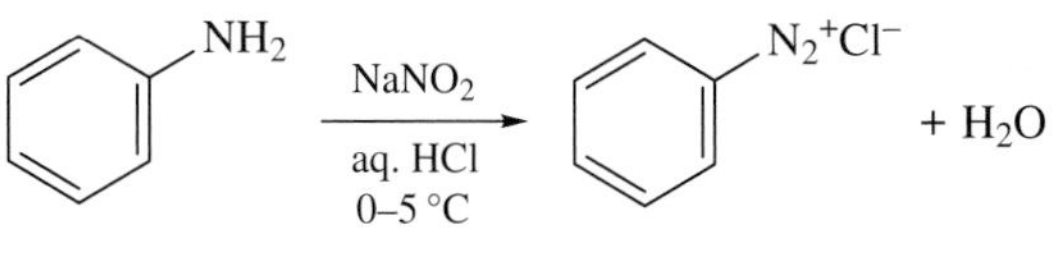

NH_2 $\xrightarrow[\text{aq. HCl, 0–5 °C}]{NaNO_2}$ $N_2^+Cl^-$ + H_2O

Scheme 8.11

The mechanism of diazotization involves generation of the electrophilic nitrosonium cation (NO^+), which nitrosates the nucleophilic amine at the nitrogen atom (Scheme 8.12). Loss of water after a series of prototropic shifts produces the diazonium salt.

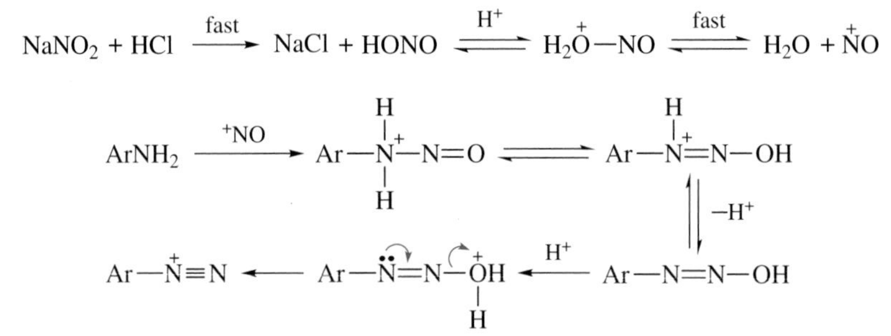

$$NaNO_2 + HCl \xrightarrow{\text{fast}} NaCl + HONO \overset{H^+}{\rightleftharpoons} H_2\overset{+}{O}{-}NO \overset{\text{fast}}{\rightleftharpoons} H_2O + \overset{+}{N}O$$

$$ArNH_2 \xrightarrow{^+NO} Ar{-}\overset{+}{N}H_2{-}N{=}O \rightleftharpoons Ar{-}\overset{+}{N}H{=}N{-}OH \overset{-H^+}{\rightleftharpoons} Ar{-}N{=}N{-}OH \xrightarrow{H^+} Ar{-}\ddot{N}{=}N{-}\overset{+}{O}H_2 \longrightarrow Ar{-}\overset{+}{N}{\equiv}N$$

Scheme 8.12

The conditions for successful diazotization depend upon the basicity of the amino group. Relatively highly basic amines such as aniline and the toluidines dissolve in aqueous hydrochloric acid. Treatment with aqueous sodium nitrite solution at 0–5 °C then very rapidly converts the amino group into the diazonium compound. Addition of nitrite is continued until there is a slight excess of nitrous acid, which is indicated by an instant dark blue colour with potassium iodide/starch paper.

The synthesis of diazonium salts of less basic amines does not proceed satisfactorily under the above conditions because of the reduced nucleophilic nature of the amino group and the reaction is usually carried out in concentrated sulfuric acid. The addition of sodium nitrite to concentrated sulfuric acid produces the stable nitrosylsulfuric acid, ($NOHSO_4$). Diazotization of the most weakly basic amines is carried out using nitrosylsulfuric acid in a mixture of one part of propionic acid in five parts of acetic acid at 0–5 °C. The propionic acid prevents the mixture from freezing.

The diazonium group is one of the most versatile functional groups in aromatic chemistry, a feature that is a consequence of the presence of a stable leaving group, a nitrogen molecule. Solid diazonium salts can lose nitrogen in an explosive manner and it is dangerous to prepare them in this state. The salts also decompose gently in solution above about 10 °C.

Aliphatic amines can also be diazotized, but the products are too unstable to be isolated and rapidly evolve nitrogen gas. The relative stability of aromatic diazonium salts is a result of delocalization of the positive charge on nitrogen into the π-system of the ring, as illustrated by the canonical forms **16–18**.

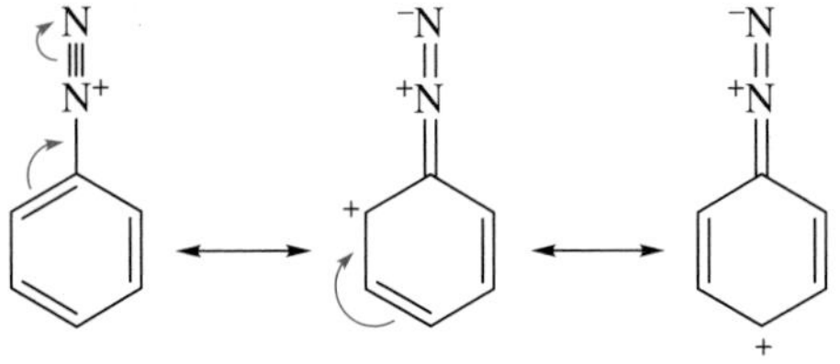

Diazonium salts can be stabilized to some extent by conversion to so-called double salts. The easiest to form are those with zinc chloride. Species such as $[ArN_2^+]_2[ZnCl_4^{2-}]$ are reasonably stable in solution, especially if kept cold.

16 **17** **18**

The synthetically useful reactions of the diazonium compounds fall into two categories: those where the nitrogen atoms are eliminated and those where they are retained. However, from a mechanistic point of view, three different types of reaction have been recognized:

1. Loss of nitrogen and generation of an aryl cation in an S_N1 reaction:

$$Ar-\overset{+}{N}\equiv N \longrightarrow Ar^+ + N_2$$

2. Loss of nitrogen and generation of an aryl radical through a one-electron reduction:

$$Ar-\overset{+}{N}\equiv N \xrightarrow{e^-} Ar^\bullet + N_2$$

3. Nucleophilic attack at nitrogen:

$$Ar-\overset{+}{N}\equiv N + :Nu^- \longrightarrow Ar-N=N-Nu$$

8.6.1 Reactions in which Nitrogen is Eliminated

Replacement of the diazonium group by a variety of other functional groups is facilitated by the presence of one of chemistry's best leaving groups, the nitrogen molecule.

Replacement by Hydrogen

It can be useful to replace an amino group with hydrogen, since this can offer a route to compounds difficult to prepare by other direct methods. The most reliable means of achieving this is by conversion to the diazonium salt and subsequent reaction with phosphinic acid (hypophosphorous acid, H_3PO_2), catalysed by copper(I) salts. A free-radical mechanism is proposed, in which copper(I) ion acts as a one-electron reducing agent and initiates a chain reaction (Scheme 8.13).

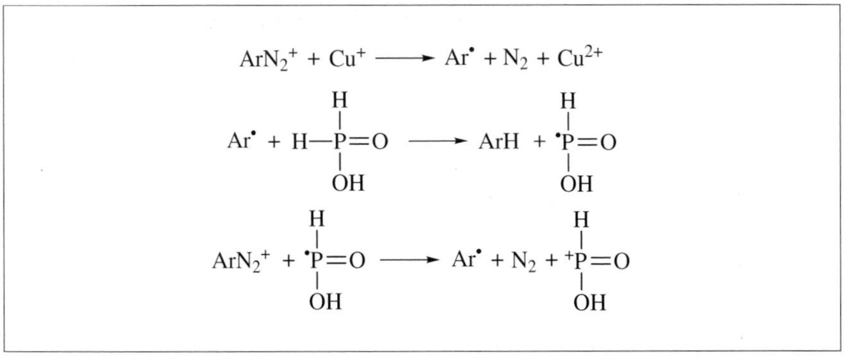

$$ArN_2^+ + Cu^+ \longrightarrow Ar^\bullet + N_2 + Cu^{2+}$$

$$Ar^\bullet + H-P(H)(OH)=O \longrightarrow ArH + {}^\bullet P(H)(OH)=O$$

$$ArN_2^+ + {}^\bullet P(H)(OH)=O \longrightarrow Ar^\bullet + N_2 + {}^+P(H)(OH)=O$$

Scheme 8.13

Diazotization of the amine in ethanol and sulfuric acid and heating the resultant solution also effects the replacement by hydrogen, but other products arising from interception of the aryl cation by different nucleophiles are also formed (Scheme 8.14).

$$ArN_2^+HSO_4^- \xrightarrow{EtOH} [Ar^+] \rightarrow ArH + MeCHO \quad ; \quad [Ar^+] \rightarrow ArOEt$$

Scheme 8.14

Replacement by a Hydroxyl Group

Phenols are formed when a diazonium salt is heated in boiling water. Nitrogen is evolved and the aryl cation reacts rapidly with water (Scheme 8.15). Since this is a nucleophilic displacement, it is preferable to use acidic conditions to ensure that no phenoxide ions are present, since this could react with unchanged diazonium salt. Sulfuric acid rather than hydrochloric acid is preferred for the diazotization to avoid trapping the highly reactive carbocation with chloride ion.

$$Ar{-}\overset{+}{N}{\equiv}N \longrightarrow Ar^+ + N_2 \xrightarrow{H_2O} ArOH$$

Scheme 8.15

Replacement by Halogen

The replacement of the diazonium group by chlorine or bromine is accomplished using the Sandmeyer reaction. Replacement with fluorine and iodine can be achieved by variations of this reaction.

In the Sandmeyer reaction, the cold diazonium salt solution is run into a solution of the copper(I) halide dissolved in the halogen acid. The complex, which usually separates, is decomposed to the aryl halide by heating the reaction mixture. The mechanism (Scheme 8.16) involves generation of an aryl radical by electron transfer from Cu(I), which then reacts with the halide ion.

$$ArN_2^+ + Cu^+ \longrightarrow Ar^\bullet + N_2 + Cu^{2+} \xrightarrow{Cl^-} ArCl$$

Scheme 8.16

Copper(I) iodide is unsatisfactory for use in the Sandmeyer reaction because of its insolubility. The iodo group is introduced by warming the diazonium salt solution in aqueous potassium iodide solution (Scheme 8.17). This method is one of the best means of introducing iodine into an aromatic ring. A one-electron reduction by the iodide ion is thought to initiate a radical reaction in a similar way to the Cu(I) ion.

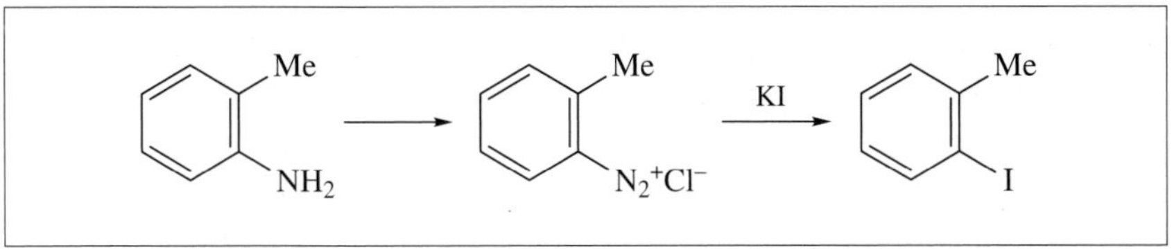

$$\text{2-MeC}_6\text{H}_4\text{NH}_2 \longrightarrow \text{2-MeC}_6\text{H}_4\text{N}_2^+\text{Cl}^- \xrightarrow{\text{KI}} \text{2-MeC}_6\text{H}_4\text{I}$$

Scheme 8.17

Introduction of a fluorine atom is achieved using the Schiemann reaction. Originally, the reaction involved gently heating the solid diazonium fluoroborate (Scheme 8.18), but improved yields result from the thermal decomposition the hexafluorophosphate, $ArN_2^+PF_6^-$, or hexafluoroantimonate, $ArN_2^+SbF_6^-$, salts.

$$ArN_2^+Cl^- + HBF_4 \longrightarrow ArN_2^+BF_4^- \xrightarrow[-N_2]{\text{heat}} Ar^+ + BF_4^- \longrightarrow ArF + BF_3$$

Scheme 8.18

Replacement by Nitrile

Introduction of the cyano group by the Sandmeyer reaction involves treatment of a diazonium solution with a solution of copper(I) cyanide in aqueous potassium cyanide (Scheme 8.19).

$$ArN_2^+Cl^- \xrightarrow[\text{aq. KCN}]{\text{Cu(I)CN}} ArCN$$

Scheme 8.19

Replacement by Aryl

Only moderate yields of biphenyls result from the Gomberg reaction, in which an acidic solution of a diazonium salt is made alkaline with sodium hydroxide solution in a two-phase mixture with an aromatic hydrocarbon such as benzene (Scheme 8.20). A variation of this reaction uses a copper catalyst. The reaction is more successful in the intramolecular version and the Pschorr reaction (discussed in Chapter 12) offers a useful route to phenanthrene.

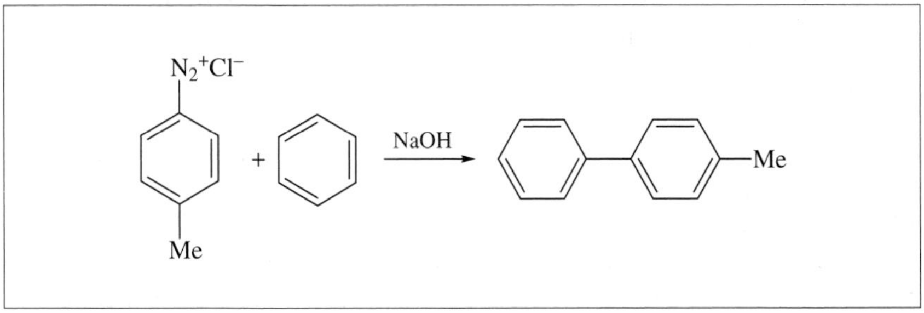

$$\text{4-MeC}_6\text{H}_4\text{N}_2^+\text{Cl}^- + \text{C}_6\text{H}_6 \xrightarrow{\text{NaOH}} \text{C}_6\text{H}_5\text{–C}_6\text{H}_4\text{–Me}$$

Scheme 8.20

Worked Problem 8.3

Q Elucidate the product that results from the treatment of a solution of 2-aminobenzophenone in hydrochloric acid with aqueous sodium nitrite solution and subsequent warming with copper powder.

A The initial reaction produces the diazonium salt from the amino group of the substituted benzophenone. In the presence of Cu, warming results in the loss of nitrogen and the generation of an aryl radical. Intramolecular attack at the *ortho* position of the adjacent phenyl ring yields the tricyclic molecule fluorenone (Scheme 8.21).

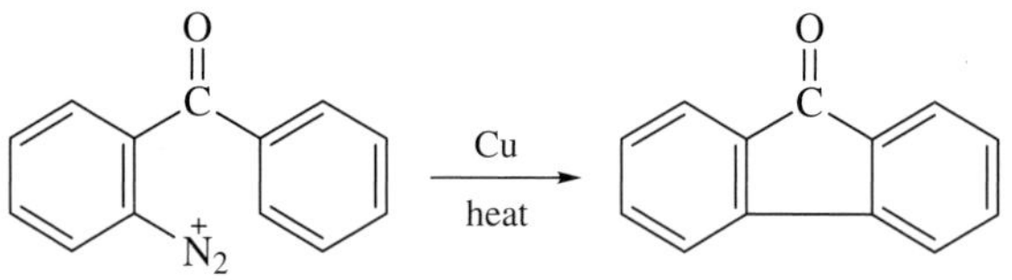

Cu, heat

Scheme 8.21

Replacement by Other Groups

Generation of an aryl radical and an aryl cation from a diazonium salt are easy processes. Both species are very reactive and are readily trapped by a wide variety of nucleophiles. Conversion of an amino group into a nitro group involves reaction of a diazonium fluoroborate with aqueous sodium nitrite solution in the presence of copper powder (Scheme 8.22).

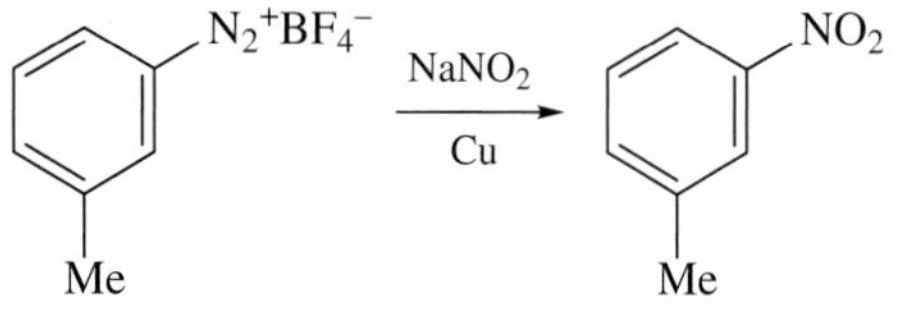

$N_2^+BF_4^-$, Me; $NaNO_2$, Cu; NO_2, Me

Scheme 8.22

The thiol (SH) group is introduced by reaction with potassium ethyl xanthate followed by acid hydrolysis. The phenylsulfanyl (phenylthio, SPh) group results from reaction with benzenethiolate ion. Sodium disulfide, Na_2S_2, yields diaryl disulfides. The arsonic acid group is introduced using **Bart's reaction**, in which a diazonium salt is reacted with sodium arsenite in the presence of a Cu(II) salt (Scheme 8.23).

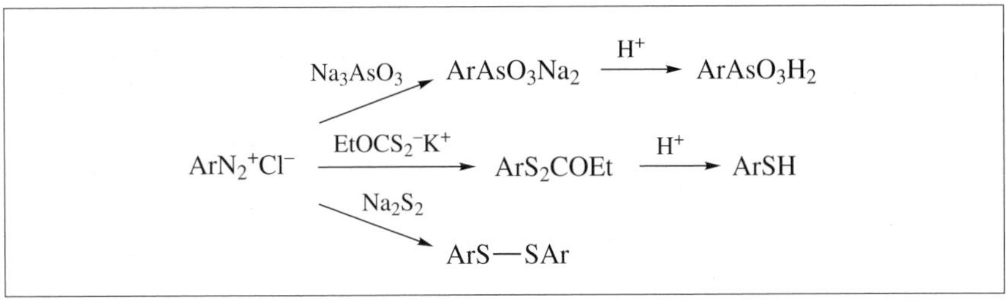

$$ArN_2^+Cl^- \xrightarrow{Na_3AsO_3} ArAsO_3Na_2 \xrightarrow{H^+} ArAsO_3H_2$$

$$ArN_2^+Cl^- \xrightarrow{EtOCS_2^-K^+} ArS_2COEt \xrightarrow{H^+} ArSH$$

$$ArN_2^+Cl^- \xrightarrow{Na_2S_2} ArS{-}SAr$$

Scheme 8.23

8.6.2 Reactions with Retention of Nitrogen

Reduction to Hydrazines

Aryl hydrazines are precursors of nitrogen heterocycles such as pyrazolones, which are useful coupling reagents.

Aryl hydrazines are formed by the reduction of diazonium salts with tin(II) chloride in hydrochloric acid (Scheme 8.24). The reduction can also be achieved by treatment with sodium sulfite solution.

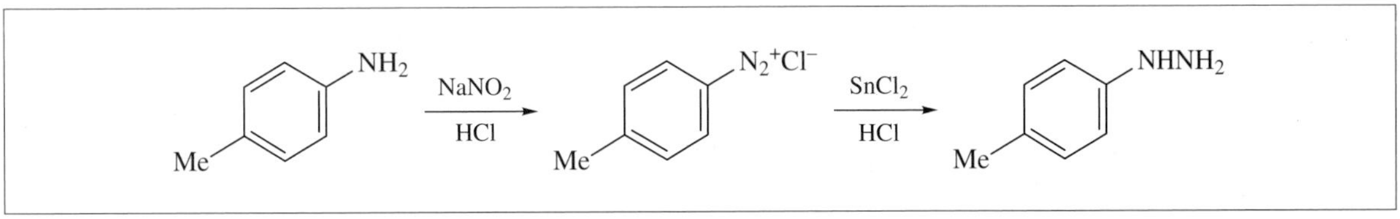

$$4\text{-}MeC_6H_4NH_2 \xrightarrow[HCl]{NaNO_2} 4\text{-}MeC_6H_4N_2^+Cl^- \xrightarrow[HCl]{SnCl_2} 4\text{-}MeC_6H_4NHNH_2$$

Scheme 8.24

Coupling Reactions

A common use of diazonium salts is in the synthesis of azo dyes. In dyestuffs chemistry, the amine that is diazotized is referred to as the diazonium component and the compound it reacts with is known as the coupling component. The whole reaction is known as azo coupling. About 50% of all commercial dyestuffs, of which two examples are shown in **19** and **20**, contain the azo group.

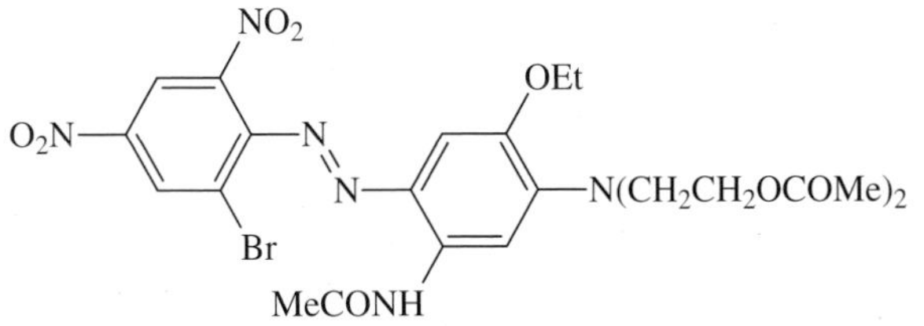

NO₂, O₂N, Br, OEt, N(CH₂CH₂OCOMe)₂, MeCONH

19 Disperse Blue 79

O, Me, NH, HO

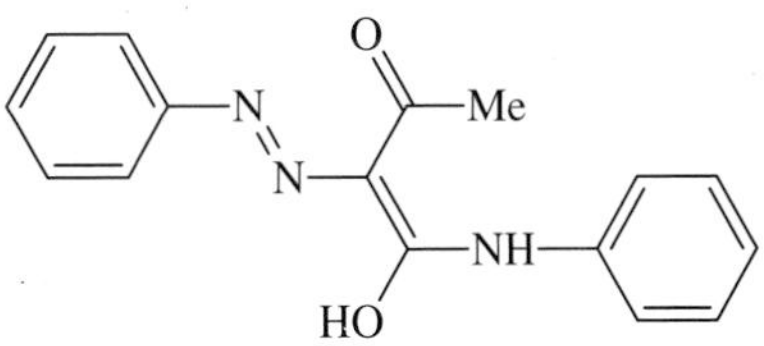

20 Disperse Yellow 16

Diazonium salts are relatively weak electrophiles. Consequently, they only react with aromatic systems which are strongly activated by the presence of powerful electron-donating groups or compounds containing

Pyridones and pyrazolones, tautomers of the hydroxypyridines and hydroxypyrazoles, respectively, are phenol-like in behaviour and are sufficiently activated to couple with diazonium salts, giving commercially valuable azo dyes.

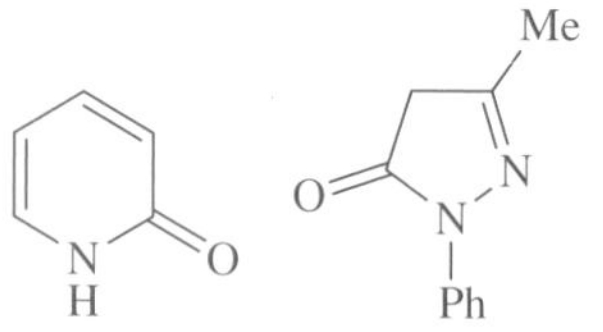

2-Pyridone A 5-pyrazolone

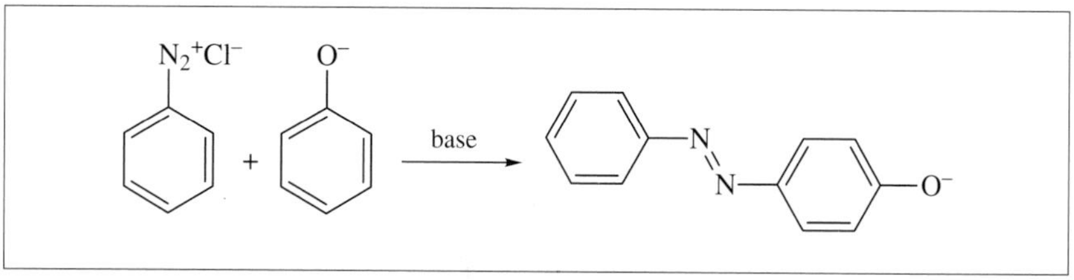

an "active" methylene group. The most common coupling components are phenols, naphthols, dialkylanilines, pyrazolones and pyridones.

Reaction with phenols and naphthols are usually carried out in the pH range 8–11, when the coupling species is the phenoxide ion. A cold, acidic solution of the diazonium salt is added to an alkaline solution of the phenol, when a fast electrophilic aromatic substitution occurs at the 4-position (Scheme 8.25). If this position is already occupied, attack occurs at the 2-position. 2-Naphthol couples at the 1-position.

Scheme 8.25

Tertiary amines react in a similar manner over the pH range 4–10. An example is the reaction between diazotized aniline and *N,N*-diethylaniline (Scheme 8.26).

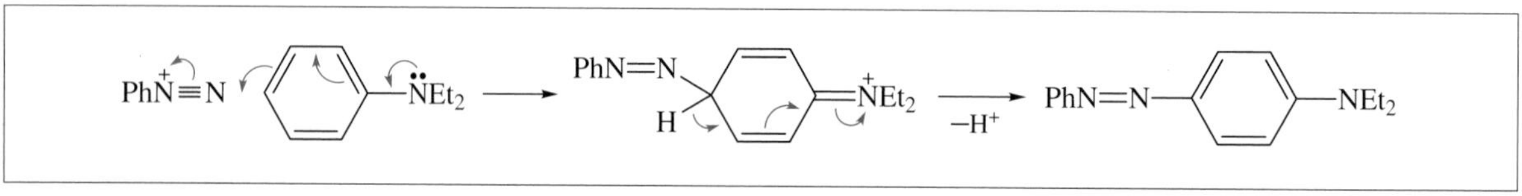

Scheme 8.26

However, primary and secondary amines usually react at the nitrogen atom rather than at carbon. Aniline, for instance, gives 4-aminoazobenzene [(4-aminophenyl)phenyldiazene] in acetate-buffered hydrochloric acid solution (Scheme 8.27). In strongly acidic solution, the *N*-coupled products rearrange to the *C*-coupled compound. Coupling conditions can be chosen which allow the *C*-coupled product to be obtained directly.

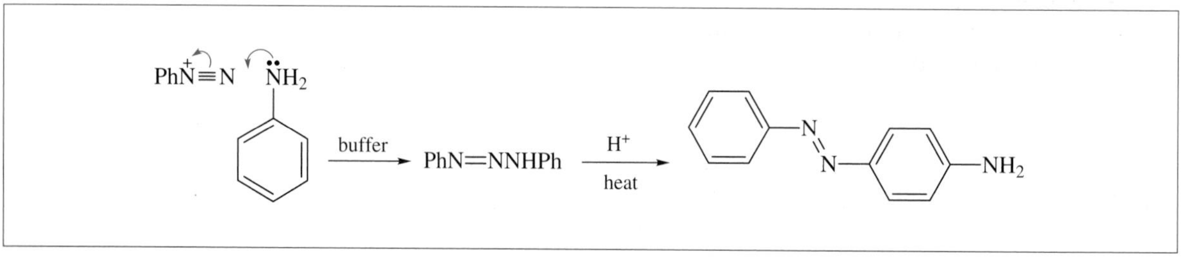

Scheme 8.27

Summary of Key Points

1. Primary aromatic amines are synthesized by reduction of nitro compounds.

2. Molecular rearrangement of the compounds ArC(=O)N–Y leads to amines.

3. The amino group is a strong electron donor and directs electrophilic attack to the *ortho* and *para* positions.

4. The amino group is readily acylated and alkylated.

5. Aromatic amines are less basic than their aliphatic analogues. Electron-withdrawing groups reduce the basicity still further.

6. Treatment of primary aromatic amines with aqueous $NaNO_2$ and acid at below 5 °C produces diazonium salts.

7. Diazonium salts are valuable reagents for the synthesis of aryl halides, nitriles, phenols, hydrazines and azo compounds.

Problems

8.1. Place the following compounds in order of increasing basicity, explaining your chosen sequence: (a) aniline, benzylamine, diphenylamine; (b) 4-aminobenzaldehyde, 4-bromoaniline, methyl 4-aminobenzoate.

8.2. Suggest a synthesis of the following compounds from the indicated starting material: (a) 1,3,5-trichlorobenzene from aniline; (b) 4-chlorobenzoic acid from aniline; (c) 1-chloro-4-nitrobenzene from aniline; (d) 1-bromo-3-fluorobenzene from benzene; (e) 3-chloroacetanilide from benzene.

8.3. Give the structures of the products from the reaction of 4-bromoaniline with the following reagents: (a) HCl; (b) $NaNO_2$ + H_2SO_4 followed by CuBr; (c) Ac_2O; (d) excess EtI; (e) excess Br_2 water.

8.4. Account for the formation of 1-naphthol when 2-aminobenzoic acid is diazotized and heated with furan and the product is treated with hydrochloric acid.

9
Aromatic Halogen Compounds

Aims

By the end of this chapter you should understand:

- How halogens can be introduced into aromatic molecules
- The reactions of aryl halides

9.1 Introduction

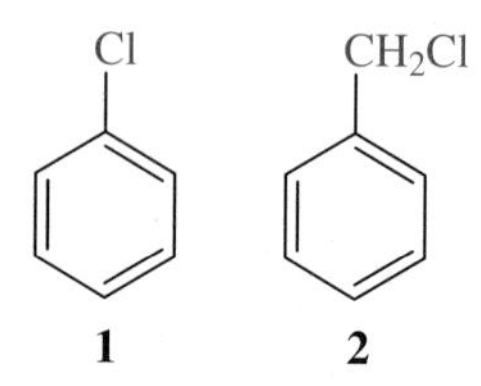

There are two distinct classes of aromatic halogen compounds. **Aryl halides** are those compounds where the halogen is directly bonded to the aromatic ring, as in chlorobenzene (**1**). Compounds where the halogen is present in an aliphatic side-chain, the **benzylic halides** such as benzyl chloride [(chloromethyl)benzene, **2**], will also be considered. The reactivity of these two categories of halogen compounds are quite different, because they are bonded to sp^2 and sp^3 hybridized carbon atoms, respectively. As we saw in Chapter 2, the aryl halides are significantly less reactive towards nucleophilic substitution than the benzylic halides. The latter resemble alkyl halides or, even more, allylic halides (CH_2=$CHCH_2X$) in their reactions.

Under certain conditions, benzene can react with halogens by **addition** rather than by substitution. In the presence of sunlight, a **free-radical reaction** takes place with chlorine that leads to addition products in which the aromatic character has been lost. The final product is hexachlorocyclohexane (benzene hexachloride), which can exist in eight possible stereoisomeric forms. The process starts with the photolytic dissociation of chlorine. Free-radical addition to the π-electron system of the aromatic ring follows and a chain reaction ensues (Scheme 9.1).

Free radicals are atoms or molecular species which contain an unpaired electron. They are usually very reactive, seeking another electron with which to pair up.

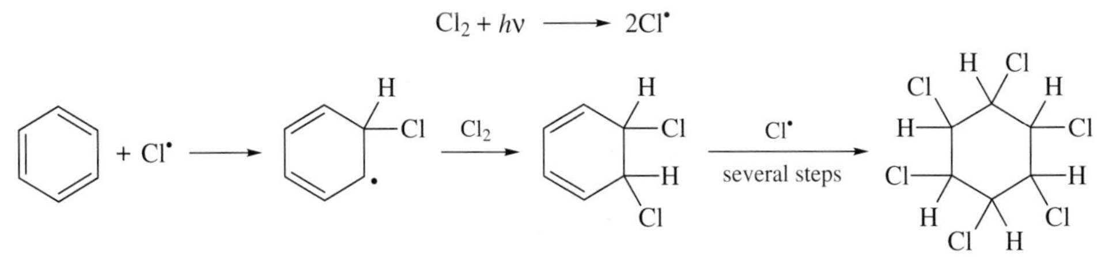

$$Cl_2 + h\nu \longrightarrow 2Cl^\bullet$$

Scheme 9.1

9.2 Synthesis of Aryl Halides

9.2.1 Direct Halogenation

The differing chemical nature and the significantly different reactivity of the halogen molecules means that various methods are necessary for their direct introduction into the benzene ring. Their reactivity increases in the order $I_2 < Br_2 < Cl_2 < F_2$. Fluorine is too reactive to allow its direct introduction, reacting explosively with benzene, and indirect methods have to be used (see Chapter 8). In practice, bromination and chlorination can be achieved effectively at moderate temperatures in the presence of a Lewis acid catalyst. The role of the Lewis acid, typically iron(III) chloride or aluminium chloride for chlorination, is to increase the electrophilic nature of the halogen through its polarization and conversion into a complex which then reacts with the π-system of the aromatic ring. Subsequent deprotonation regenerates both the aromatic π-system and the Lewis acid, which is therefore truly catalytic in behaviour (Scheme 9.2).

Remember that a Lewis acid is a species that accepts an electron pair.

The most powerful chlorinating agent is claimed to be the Silberrad reagent, formed from disulfur dichloride, S_2Cl_2, sulfuryl chloride, SO_2Cl_2, and aluminium chloride.

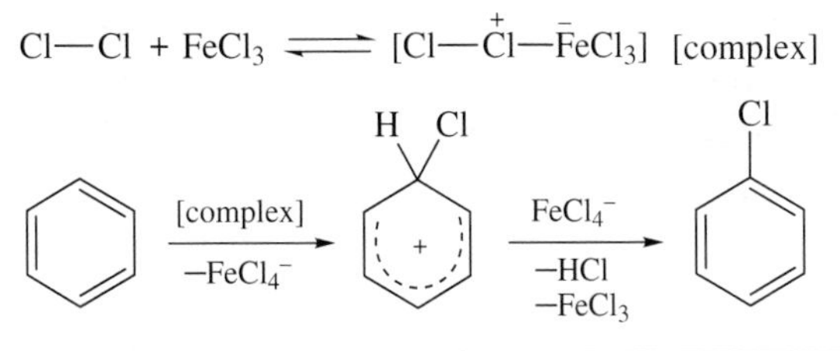

$$Cl{-}Cl + FeCl_3 \rightleftharpoons [Cl{-}\overset{+}{Cl}{-}\overset{-}{Fe}Cl_3] \text{ [complex]}$$

Scheme 9.2

Worked Problem 9.1

Q Bromobenzene is readily synthesized from benzene by reaction with bromine in the presence of pyridine. Suggest a mechanism for this route.

A Pyridine is behaving as a Lewis base, reacting with bromine at the nitrogen atom to generate *N*-bromopyridinium bromide. This acts as a source of electron-deficient bromine, $Br^{\delta+}$, rather than a full Br^+ cation, which reacts with the benzene ring as shown in Scheme 9.3.

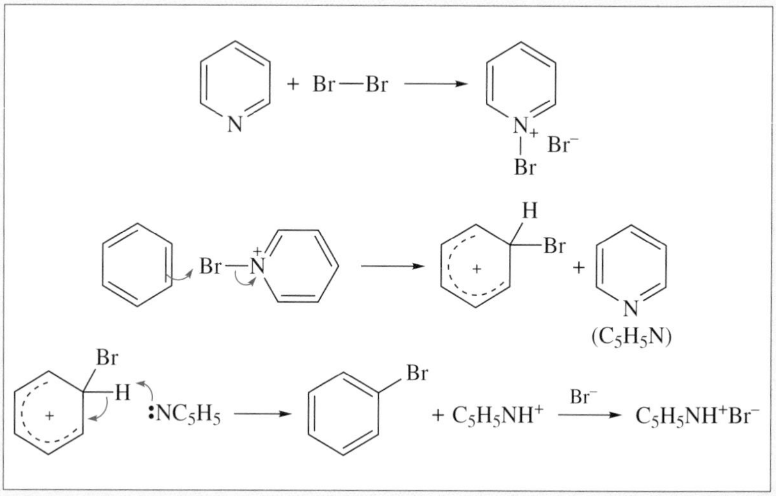

+ Br—Br → N^+ Br^- Br

H Br + N (C_5H_5N)

Br H :NC_5H_5 → Br + $C_5H_5NH^+$ $\xrightarrow{Br^-}$ $C_5H_5NH^+Br^-$

Scheme 9.3

Both trifluoroacetyl hypoiodite (CF_3CO_2I) and iodine in peroxyacetic acid, $MeCO_3H$, can also be used for direct iodination. The use of silver sulfate and iodine in sulfuric acid as an iodinating mixture is based on the insolubility of AgI. Iodide ion, I^-, is thus removed from solution, leaving the electrophilic iodonium ion, I^+.

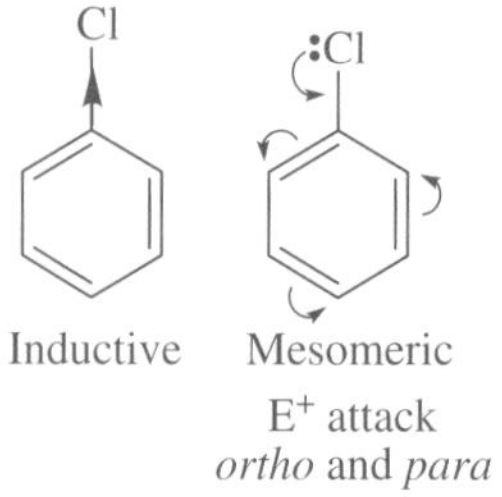

Cl :Cl

Inductive Mesomeric
E^+ attack
ortho and *para*

Monohalogenation of these reactive compounds can be achieved either by prior conversion of the activating group into a less reactive derivative, or by using specialized halogenating reagents.

Greater activation is needed to achieve direct iodination because of the lower reactivity of iodine. Oxidizing agents such as nitric acid and hydrogen peroxide convert the halogen into a more reactive electrophilic species. The I_2/HNO_3 mixture, perhaps containing HNO_2I^+, will iodinate benzene directly.

Increasing the reaction temperature and the amount of halogen used results in the introduction of further halogen atoms. It was noted in Chapter 2 that the halogen atom in chlorobenzene is *ortho*/*para* directing, but deactivating to electrophilic substitution as a result of opposing mesomeric (+M) and inductive (–I) effects. Consequently, disubstitution leads to a mixture of the 1,2- and, mainly, 1,4-isomers under conditions similar to those required to attack benzene.

Aromatic systems substituted with electron-donating groups are more readily halogenated than benzene. Consequently, other synthetic routes or reagents are sometimes used to avoid polyhalogenation and the formation of isomeric mixtures. For example, the iodination of toluene gives a mixture of 2- and 4-iodotoluenes; each isomer can be prepared individually from the appropriate toluidine *via* the diazonium salt (see Chapter 8).

In the case of aromatic systems activated towards electrophilic attack by strongly electron-donating substituents, a catalyst may not be required. For instance, both phenol and aniline are tribrominated by bromine at room temperature. Even iodine attacks aniline in the presence of only a scavenger for the liberated hydrogen iodide.

Worked Problem 9.2

Q How would you synthesize 4-bromoaniline from aniline?

A Direct bromination of aniline gives 2,4,6-tribromoaniline, so it is necessary to reduce the electron-donating power of the amino group in order to moderate the reaction with electrophiles. Acetylation of aniline with acetic anhydride leads to acetanilide. The substituted amide group, NHCOMe, is less electron donating than a free amino group, because the carbonyl group competes effectively with the aromatic ring for the lone pair of electrons on the N atom, as described in Chapter 8.

Bromination in acetic acid now results in monobromination, but it gives a mixture of isomers. Fortunately, steric hindrance by the bulky acetamido group shields the *ortho* positions from attack to a large extent; only about 10% of the 2-bromo derivative is formed. This isomer remains in solution when the mixed product is recrystallized and pure 4-bromoacetanilide is readily isolated. Finally, hydrolysis of the amide function liberates the amino group and 4-bromoaniline results (Scheme 9.4).

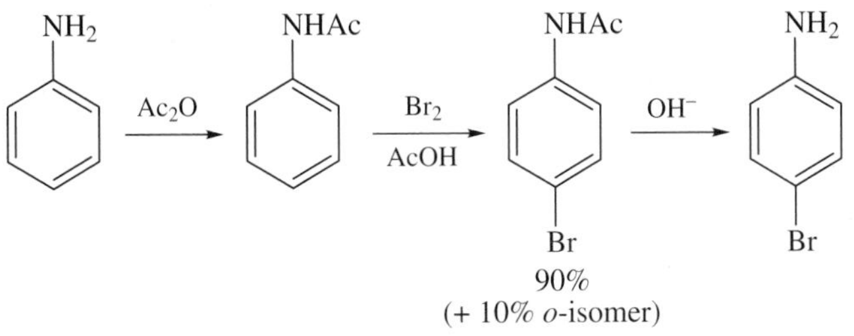

NH_2 —(Ac_2O)→ NHAc —(Br_2, AcOH)→ NHAc / Br 90% (+ 10% *o*-isomer) —(OH^-)→ NH_2 / Br

Scheme 9.4

9.2.2 From Amines

As described in Chapter 8, the indirect replacement of an amino group by a halogen *via* diazotization of a primary aromatic amine is a valuable route to aryl halides. The chemistry of these processes covering the introduction of all the main halogens can be found in that chapter.

9.2.3 From Carboxylic Acids

The Hunsdiecker reaction, in which the silver salt of a benzoic acid is thermally decarboxylated in the presence of bromine, gives moderate yields of aryl bromides. The mechanism is uncertain, but may involve the generation of aryl radicals (Scheme 9.5).

$$ArCO_2Ag + Br_2 \longrightarrow ArCO_2Br + AgBr$$

$$ArCO_2Br \longrightarrow Br^{\bullet} + ArCO_2^{\bullet} \longrightarrow Ar^{\bullet} + CO_2$$

$$Ar^{\bullet} + Br_2 \longrightarrow ArBr + Br^{\bullet}$$

$$Ar^{\bullet} + ArCO_2Br \longrightarrow ArBr + ArCO_2^{\bullet}$$

Scheme 9.5

9.2.4 Commercial Synthesis of Chlorobenzene

Chlorobenzene is an important commercial solvent, although it is less used nowadays because of environmental concerns. It is produced commercially by the **Raschig process**, in which a mixture of benzene vapour, air and hydrogen chloride is passed over a copper chloride catalyst.

9.3 Reactions of Aryl Halides

This section is concerned with reactions of the *C*-halogen group itself. It is generally true that aryl halides are less reactive than alkyl halides, but the former compounds do undergo a number of useful reactions.

An important example of this behaviour is provided by the reaction of 1-fluoro-2,4-dinitrobenzene (**3**) with the terminal amino group of proteins. Subsequent acidic hydrolysis yields the yellow 2,4-dinitrophenyl derivative of the terminal amino acid of the protein, which can then be identified. With the help of this technique of end-group analysis, Sanger was able to determine the primary structure of insulin.

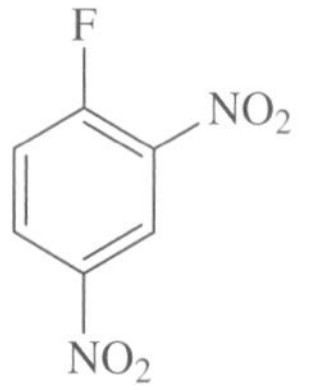

9.3.1 Nucleophilic Substitution

Nucleophilic substitution of aryl halogen atoms requires significant energy input. Thus, in the **Dow process** for the synthesis of phenol from chlorobenzene, the chlorine atom is only successfully hydrolysed by aqueous sodium hydroxide at 300 °C under pressure. Displacement by ammonia is achieved at 200 °C over copper(I) oxide and conversion to benzonitrile occurs using copper(I) cyanide in boiling dimethylformamide, $HCONMe_2$.

However, nucleophilic substitution is helped by the presence of electron-withdrawing groups in the molecule. Consequently, 1-chloro-4-nitrobenzene is hydrolysed to 4-nitrophenol at 200 °C and 4-nitroaniline can be produced using ammonia in ethanol at 150 °C. The presence of two nitro groups further activates the halogen and 1-chloro-2,4-dinitrobenzene reacts easily with a variety of nucleophiles.

F
NO_2
NO_2
3

Not surprisingly, three nitro groups have an even greater influence on the reactivity of the halogen and 1-chloro-2,4,6-trinitrobenzene (picryl chloride) is hydrolysed to 2,4,6-trinitrophenol (picric acid) in boiling water. The trivial names tell us that this aryl halide behaves as an acyl halide and the phenol as an acid.

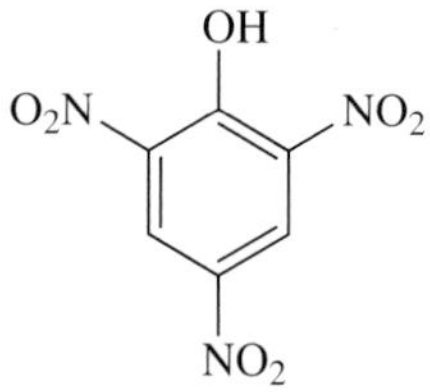

OH
O_2N
NO_2
NO_2
picric acid

A nitro group deactivates an aromatic ring to electrophilic attack, but it activates the ring towards nucleophilic substitution.

The electron-withdrawing nature of the halogen is supported by the powerful effect of an *ortho*- and/or a *para*-nitro group, so that the carbon atom to which the halogen is attached is sufficiently electron deficient to promote attack by the nucleophile. The resulting intermediate is stabilized by the nitro group(s). Elimination of halide ion restores the aromatic system and completes the bimolecular process, which overall is a further example of an addition–elimination sequence (Scheme 9.6). It is emphasized that electron-withdrawing groups other than a nitro group exert similar activating effects towards nucleophilic attack.

The resonance interaction involving a nitro group in the 2- and 4-positions and the carbon atom to which the halogen is attached is not possible when the nitro group is in the 3-position.

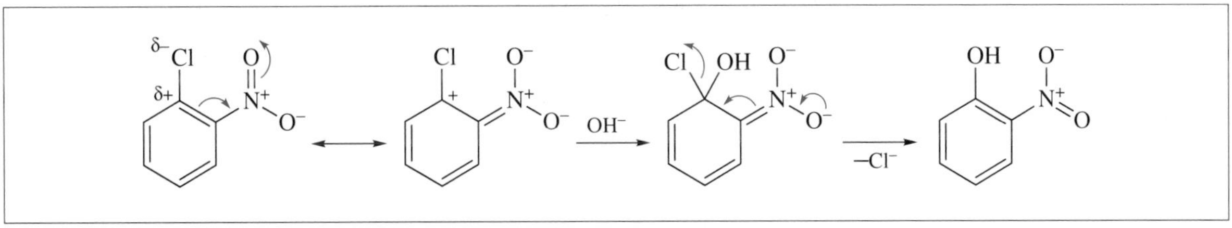

Scheme 9.6

When chlorobenzene is treated with the strong base sodamide, $NaNH_2$, aniline is formed. This reaction is more complex than it appears, since if the chlorobenzene is labelled at C-1 with the isotope ^{14}C (indicated by an asterisk), then the product consists of equal amounts of aniline with the label at C-1 and C-2. The reaction proceeds by abstraction of the weakly acidic hydrogen atom on the carbon atom next to the C–halogen group by the strong base, NH_2^-, followed by loss of halide ion. Ammonia then adds to the apparent triple bond of the resulting reactive intermediate, **benzyne (1,2-dehydrobenzene)**. The mechanism involves an elimination–addition sequence. However, the entering group does not necessarily occupy the same position as the leaving group. The process is sometimes referred to as ***cine* substitution** (Scheme 9.7).

Molecules that lack a hydrogen atom adjacent to the halogen, such as 2-bromo-1,3,5-trimethylbenzene, do not take part in this type of reaction.

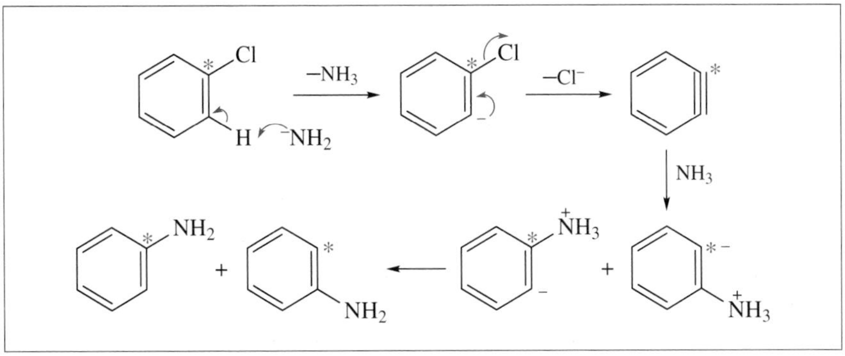

Scheme 9.7

Worked Problem 9.3

Q When labelled fluorobenzene is treated with 2 moles of phenyllithium, followed by acidification on completion of the reaction, biphenyl is formed in which only 50% of the product has the aryl–aryl bond at the labelled carbon atom. Explain.

A Phenyllithium is polarized such that the phenyl group is carbanionic in character and behaves as a strong base, abstracting the 2-H of fluorobenzene. The resulting *ortho* lithiation is followed by loss of lithium fluoride to form the benzyne intermediate. The second molecule of phenyllithium adds across the triple bond in a non-selective manner to generate biphenyl with 50% of the product with the aryl–aryl bond at the original labelled carbon atom, according to the mechanism in Scheme 9.8.

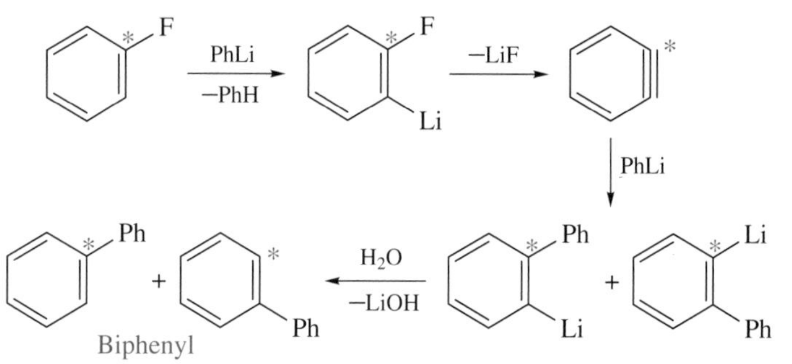

PhLi, −PhH; −LiF; PhLi; H_2O, −LiOH

Biphenyl

Scheme 9.8

A Diels–Alder reaction is the formation of a cyclic compound from a diene and an electron-deficient alkene (a dienophile). Benzyne acts as the dienophile.

Evidence for the benzyne intermediate is extensive. The most compelling is its capture in a **Diels–Alder reaction** by a diene such as furan, illustrated by the reaction of 1-bromo-2-fluorobenzene with lithium amalgam in the presence of furan (Scheme 9.9).

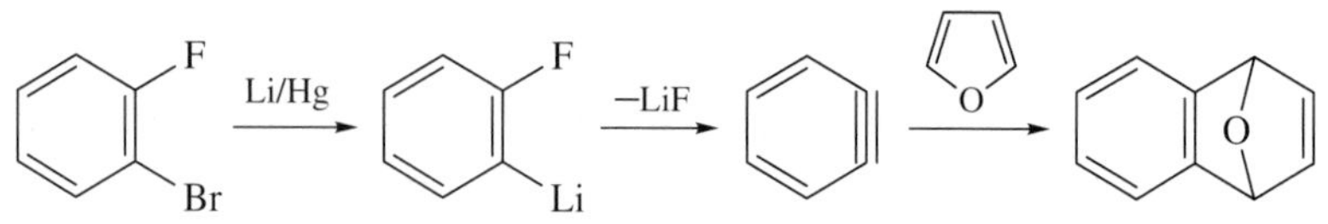

Li/Hg; −LiF

Scheme 9.9

Probably the most important reaction of aryl halides is the formation of **Grignard reagents** and **aryllithium compounds**, both of which are widely used in synthesis. These reactions are covered in Chapter 10.

9.3.2 Reactions at Ring Carbon Atoms

Halogen atoms such as chlorine and bromine deactivate the ring to electrophilic substitution by the inductive (–I) effect, but the mesomeric (+M) donation of electrons directs substitution to the *ortho* and *para* positions.

9.4 Aromatic Halogen Compounds Substituted in the Side Chain

In the absence of a Lewis acid, halogenation of toluene at its boiling point with bromine or chlorine and under UV irradiation (*e.g.* sunlight) occurs in the side chain. The reaction proceeds by a free-radical mechanism that is initiated by the photolytic dissociation of a chlorine molecule (Scheme 9.10). The benzyl radical is stabilized by resonance (see Chapter 3).

Chlorination of the side chain can also be achieved using sulfuryl chloride in the dark, but in the presence of a radical initiator such as benzoyl peroxide, and using *t*-butyl hypochlorite, $Me_3C–OCl$, from which free radicals are generated on gentle warming.

The trivial names for these aryl alkyl halides are quite confusing and we have chosen to use systematic names to help you.

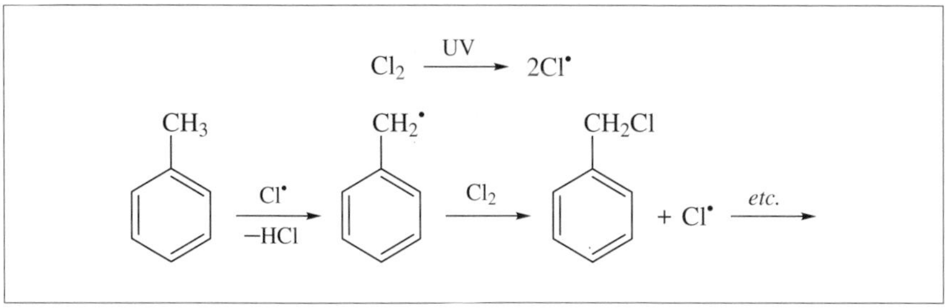

$$Cl_2 \xrightarrow{UV} 2Cl^{\bullet}$$

$$C_6H_5CH_3 \xrightarrow[-HCl]{Cl^{\bullet}} C_6H_5CH_2^{\bullet} \xrightarrow{Cl_2} C_6H_5CH_2Cl + Cl^{\bullet} \xrightarrow{etc.}$$

Scheme 9.10

It is possible to replace all three hydrogen atoms of the methyl group of toluene sequentially by chlorine leading to (chloromethyl)benzene (**4**), (dichloromethyl)benzene (**5**) and (trichloromethyl)benzene (**6**). Introduction of the first chlorine atom proceeds at a much faster rate than the second and so it is possible to prepare (chloromethyl)benzene selectively. In order to achieve the required degree of chlorination, chlorine gas is passed into the reaction mixture until the mass gain corresponds to the appropriate level of substitution.

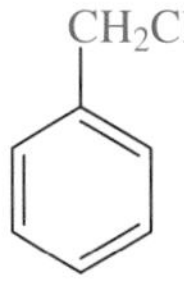

The higher homologues of toluene such as ethylbenzene are not usually halogenated selectively and mixtures are often produced (see Chapter 3). In the case of ethylbenzene itself, the major product of chlorination is the 1-substituted product (56%). Bromine is more selective and the 1-bromo derivative **7** is formed exclusively.

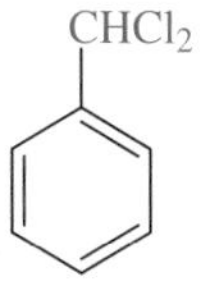

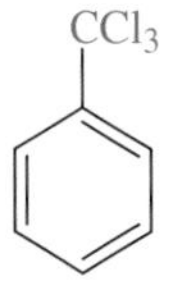

N-Bromosuccinimide (**8**) can also be used to bring about side-chain bromination, toluene yielding (bromomethyl)benzene for example.

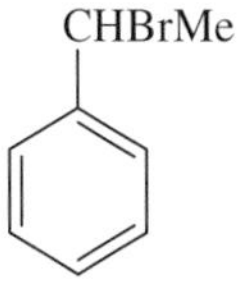

(Chloromethyl)benzene behaves like an alkyl halide towards nucleophiles, although it is more reactive than alkyl halides in both S_N1 and S_N2 reactions. In the former case, the intermediate carbocation is

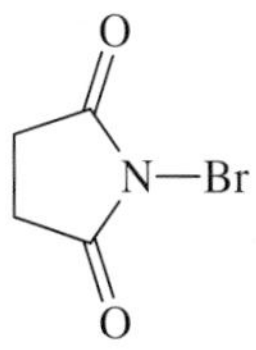

$C_6H_5CH_2Cl$ **4**; $C_6H_5CHCl_2$ **5**; $C_6H_5CCl_3$ **6**; $C_6H_5CHBrMe$ **7**; N-bromosuccinimide (N—Br) **8**

Greater stabilization can be achieved through the attachment of a second phenyl ring, producing the diphenylmethyl cation, Ph_2CH^+. A third ring gives the triphenylmethyl cation, Ph_3C^+, which is even more stable. The triphenylmethane dyes, *e.g.* Crystal Violet (below), are derived from the latter species and owe their colour to the extensive delocalization of the cationic charge, essentially on the terminal nitrogen atoms.

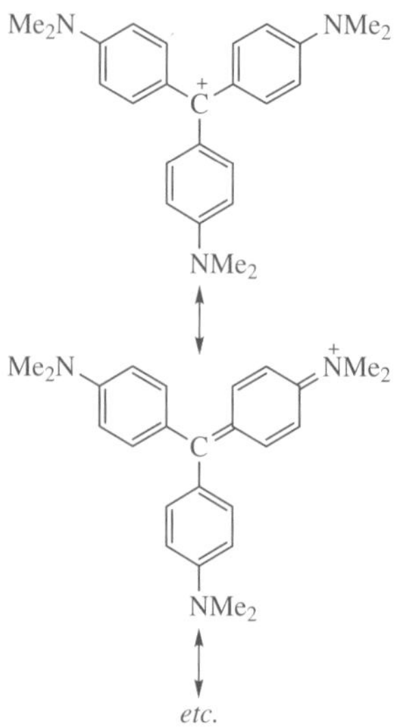

stabilized by resonance with the aromatic ring (Scheme 9.11). Hydrolysis of these side-chain halides is generally easy, especially if there is an electron-withdrawing substituent in the ring. Many are lachrymatory, contact of the vapour with moisture on the eye resulting in hydrolysis and liberation of HX. This irritates the eye and tears are produced. An industrial route to benzaldehyde from toluene involves hydrolysis of the intermediate product $PhCHCl_2$.

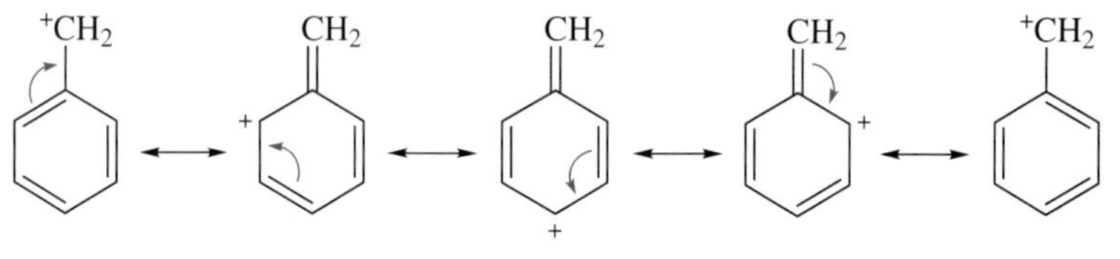

Scheme 9.11

Summary of Key Points

1. Aryl bromides and chlorides can be synthesized by direct reaction of an aromatic compound with the halogen in the presence of a Lewis acid.

2. Aryl fluorides and iodides are best prepared from diazonium salts.

3. Although a halogen atom is electron withdrawing, it directs electrophilic attack to the 2- and 4-positions.

4. Chlorination of benzene under free-radical conditions results in addition of chlorine and destruction of the aromatic system.

5. Free-radical chlorination of alkyl derivatives of benzene occurs at the alkyl group.

6. Nucleophilic displacement of the halogen can be achieved when electron-withdrawing groups are also present in the aromatic ring.

7. The reaction of strong bases with aryl halides results in the generation of an aryne through loss of HX. Arynes can also be formed by reaction of aryl halides with lithium and organolithium compounds.

Problems

9.1. Identify the product from each of the following reactions: (a) 1-chloro-4-nitrobenzene + sodium methoxide; (b) 1,2-dichloro-4-nitrobenzene + 1 mole of sodium methoxide; (c) 4-(2-chlorophenyl)-butanonitrile + sodium amide; (d) 3-bromo-4-methylbiphenyl + potassium amide.

9.2. Suggest routes to the following compounds starting from benzene: (a) 1-chloro-4-nitrobenzene; (b) 1-bromo-3-nitrobenzene; (c) 4-methoxybenzonitrile; (d) fluorobenzene.

9.3. Explain the following reaction sequences, identifying the unknown compounds in each case and drawing out the structures: (a) 1-chloro-2,4-dinitrobenzene + phenylhydrazine → compound X; then compound X + acetophenone → orange precipitate Z; (b) bromobenzene + Ac_2O + $AlCl_3$ → mainly compound A; then compound A + NH_2OH → compound B; then compound B + H_2SO_4 → compound D; (c) the peptide $H_2NCH(Ph)CONHCH(Me)CO$–polypeptide is treated with 1-fluoro-2,4-dinitrobenzene and the peptide bonds in the product are hydrolysed, giving a mixture containing many amino acids and a yellow compound, G.

10
Organometallic Reactions

Aims

By the end of this chapter you should understand:

- The ways in which metals can be used in aromatic chemistry
- How metal complexes modify reactivity of an aromatic compound from that which is normally observed
- The mechanisms by which these organometallic processes occur

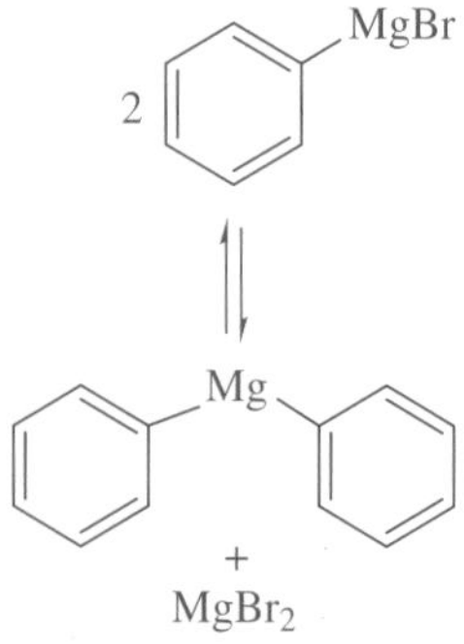

10.1 Grignard and Organolithium Reagents

10.1.1 Preparation

In solution, arylmagnesium bromides (ArMgBr) are in equilibrium with the diarylmagnesium, Ar_2Mg, and $MgBr_2$ (see below). This is known as the **Schlenk equilibrium**. The exact position of the equilibrium is solvent dependent.

2 PhMgBr ⇌ Ph_2Mg + $MgBr_2$

Aryl bromides and iodides react with magnesium and lithium to form **Grignard reagents** (ArMgX) and **aryllithium compounds** (ArLi), respectively (Scheme 10.1). The analogous chloro compounds behave similarly, but are less reactive. The normal method of preparing these highly reactive species is to add the aryl halide slowly to a stirred suspension of either magnesium turnings or finely divided lithium pieces in an anhydrous solvent such as tetrahydrofuran (THF) or diethyl ether. The species are probably formed by an electron transfer mechanism.

It is sometimes more convenient to prepare organolithium reagents by halogen–lithium exchange between an aryl halide and butyllithium (BuLi). The reaction takes place because the aryllithium is more stable than butyllithium; an sp^2 carbon is better able to stabilize a negative charge than an sp^3 carbon (Scheme 10.2). Butyllithium is available commercially in bulk quantities.

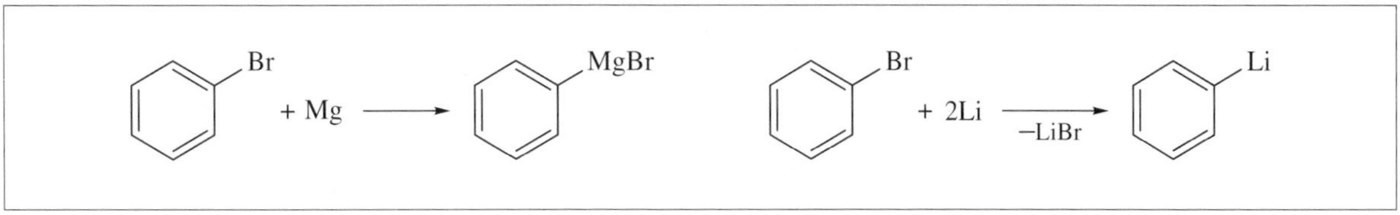

Br + Mg → MgBr

Br + 2Li →(−LiBr) Li

Scheme 10.1

PhBr ⇌(BuLi) PhLi + BuBr

Scheme 10.2

10.1.2 Directed Orthometallation

It might be expected that strongly basic alkyllithium reagents (RLi) would deprotonate an arene (ArH) directly to form the aryllithium (ArLi) and the alkane (RH). Although this reaction does occur, it is usually extremely slow and side reactions may compete. In addition, deprotonation of most substituted benzenes will probably occur in a random manner rather than at a particular position in the benzene ring.

Such deprotonations may be achieved with **Schlosser's "super base"**, a combination of butyllithium and potassium *tert*-butoxide, Me_3COK (Bu^tOK). This reagent is even more basic than organolithium species.

However, certain substituents can make the process feasible because they are able to stabilize the aryllithium. For instance, treatment of methoxybenzene with BuLi leads readily to the 2-lithio derivative **1** (Scheme 10.3).

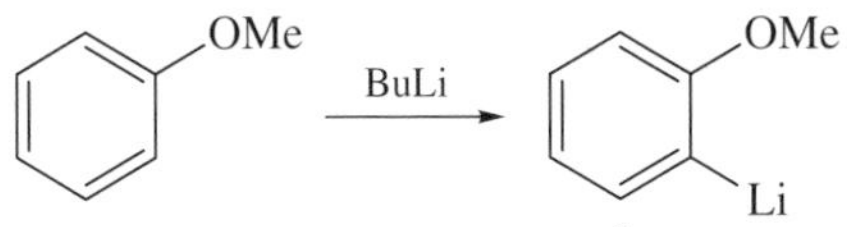

OMe →(BuLi) OMe, Li **1**

Scheme 10.3

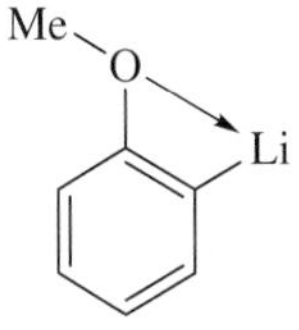

The *ortho* substituent stabilizes the molecule by coordination to the lithium as shown in **2**. This process is only possible when the lithium occupies the 2-position and so lithiation occurs exclusively at this site. Substituents that can behave in this way are known as **directing metallation groups** (DMG). The coordinating ability of a substituent varies and hence their effectiveness in directing metallation to the *ortho* position is variable. The process is known as **directed orthometallation** and is a significant development in the field of aromatic chemistry.

Me O Li **2**

Other directing groups include $CONR_2$, CH=NR, CH_2NR_2, C≡N, CF_3, $CH(OR)_2$, and the heterocyclic groups

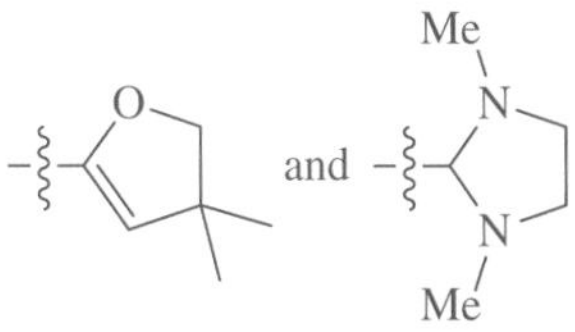

O and Me N N Me

We have seen in Chapter 2 that electrophilic attack of methoxybenzene gives a mixture of the 2- and 4-isomers. The significance of the orthometallation process is the ease with which the 2-substituted compound can be obtained exclusively through reaction of the organolithium product with a variety of electrophiles (Scheme 10.4).

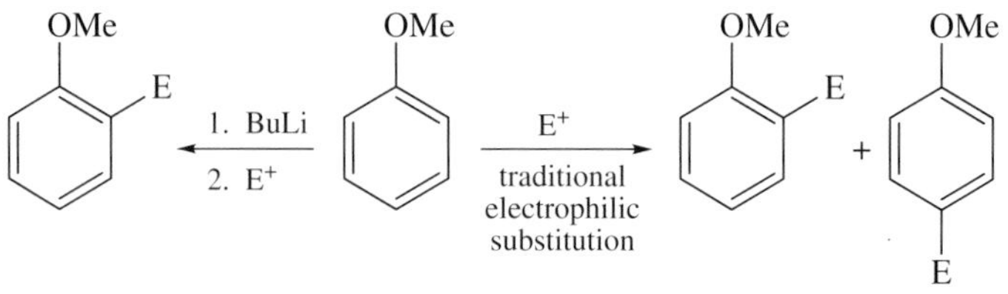

OMe, E; 1. BuLi, 2. E^+; OMe; E^+, traditional electrophilic substitution; OMe, E + OMe, E

Scheme 10.4

10.1.3 Reactions of Grignard and Aryllithium Reagents

The reactions of arylmagnesium and aryllithium compounds are similar and are analogous to the reactions of the corresponding alkyl species. These reagents are strong bases, particularly the lithium compounds, and react rapidly with weak acids such as water and alcohols to form the arene. It is therefore important when using these reagents that reactants, solvents and apparatus are dry and free of acid.

The aryl unit of these species shows carbanionic properties and consequently reacts with electrophiles. As a result, these organometallic reagents are of great value for the synthesis of a wide range of aromatic compounds. In particular, reaction with carbon electrophiles such as carbonyl compounds and nitriles results in the formation of a new carbon–carbon bond.

See Chapter 6 for other examples of carbanions reacting with carbonyl compounds and a discussion of the reaction.

Benzyl alcohols can be prepared by reaction with formaldehyde gas or more conveniently with paraformaldehyde (polymethanal) (Scheme 10.5), but reaction with other aldehydes yields **secondary alcohols**

$$ArMgBr \xrightarrow[\text{or } (CH_2O)_n]{CH_2O} ArCH_2OH \qquad ArLi \xrightarrow{RCHO} ArCH(OH)R$$

Scheme 10.5

Primary alcohols: RCH_2OH
Secondary alcohols: R_2CHOH
Tertiary alcohols: R_3COH

Reaction with the carbonyl group of a ketone yields **tertiary alcohols** (Scheme 10.6) and these can also be formed from the reaction of two equivalents of the organometallic reagent with an ester.

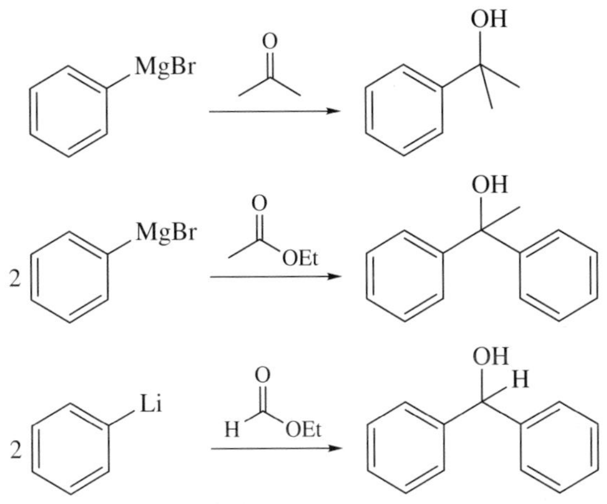

Esters also undergo reactions with one equivalent of Grignard reagent to form ketones, but only under controlled conditions and provided that the ester is not too reactive. Formation of the tertiary alcohol usually takes place with excess reagent.

Scheme 10.6

Worked Problem 10.1

Q A convenient synthesis of aldehydes involves the reaction of a Grignard reagent with triethyl orthoformate [triethoxymethane, $HC(OEt)_3$]. Suggest a mechanism for this reaction and hence explain why the reaction stops at the aldehyde stage.

A An electrophilic species is generated through the Lewis acid behaviour of the Grignard reagent. This converts the tri-alkoxymethane into a reactive oxonium ion, which is effectively an ester alkylated on oxygen. The Grignard reagent adds to this species to give an acetal, a protected aldehyde (see Chapter 6). Further reaction to the secondary alcohol is therefore prevented. The free aldehyde is liberated by hydrolysis in a subsequent step (Scheme 10.7).

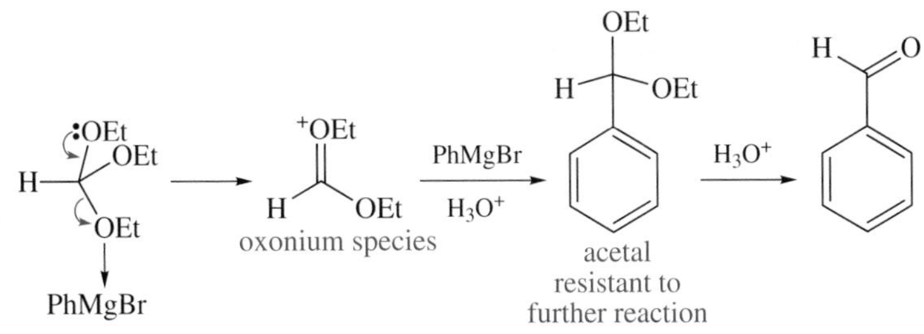

Scheme 10.7

The imine can be regarded as a protected ketone.

Ketones can be prepared by the reaction of Grignard reagents with nitriles. The initial product is a ketimine, which does not react further with the Grignard reagent, thereby preventing formation of the tertiary alcohol. Subsequent hydrolysis gives the ketone (Scheme 10.8).

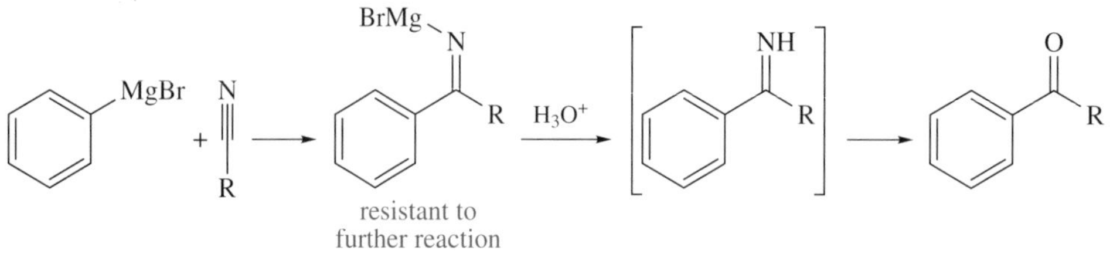

PhMgBr + N≡C–R → Ph–C(=N–MgBr)–R (resistant to further reaction) $\xrightarrow{H_3O^+}$ [Ph–C(=NH)–R] → Ph–C(=O)–R

Scheme 10.8

Aromatic carboxylic acids are prepared by reaction of a Grignard reagent with carbon dioxide, as either the solid or gas (Scheme 10.9).

$$ArMgBr \xrightarrow{CO_2} ArCO_2H$$

Scheme 10.9

10.2 Electrophilic Metallation

Salts of **mercury(II)** and **thallium(III)**, such as mercury(II) acetate, $Hg(OAc)_2$, and thallium trifluoroacetate, $Tl(OCOCF_3)_3$, are reactive electrophilic metallating species which attack benzene directly. Electron-donating groups in the aromatic ring accelerate the reaction in the conventional manner and direct attack to the *ortho* and *para* positions. In addition, metal-chelating substituents, such as amide, promote the reaction and direct attack to the 2-position in a manner similar to that seen in the directed orthometallation reaction.

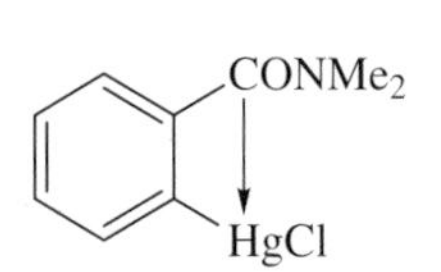

The **organomercury compounds** formed in such reactions are not of great synthetic importance, although they do undergo nitrosation reactions with nitrosyl chloride, NOCl. They can also be used as the organometallic component in certain palladium-mediated coupling processes.

Great care must be taken when using this methodology because mercury and thallium compounds are extremely toxic.

Organothallium compounds are useful because they react with potassium iodide and potassium cyanide to afford aryl iodides and nitriles, respectively (Scheme 10.10). This route offers an alternative to diazonium chemistry (see Chapter 8) for the introduction of these functional groups.

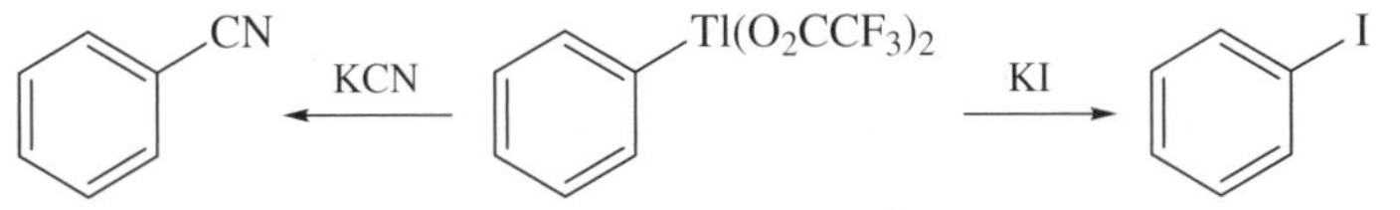

KCN ← $PhTl(O_2CCF_3)_2$ → KI
PhCN ← → PhI

Scheme 10.10

10.3 Transition Metal Mediated Processes

Transition metal complexes are widely used in aromatic chemistry and organic synthesis in general. They are of particular value because complexation of an organic molecule to a metal centre often modifies its reactivity. The metal can subsequently be removed. After discussing the ways by which these complexes react, we will discuss their use in the synthesis of aromatic compounds.

The chemistry of organometallic transition metal σ-complexes can largely be explained by the operation of fundamental processes such as oxidative addition, reductive elimination and β-elimination.

The process is termed oxidative addition because the oxidation state of the metal increases by two (as the groups A and B are normally more electronegative than the metal) and the coordination number also increases by two. These reactions may be illustrated using standard "arrow-pushing" diagrams, but here the arrows convey no real mechanistic meaning.

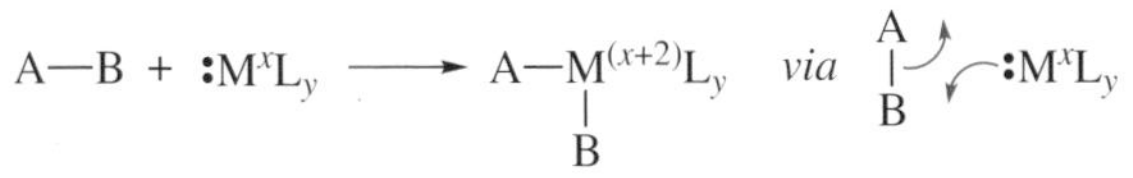

10.3.1 Oxidative Addition

Oxidative addition is the process by which a metal atom inserts into an existing bond. The metal is thus acting simultaneously as a Lewis acid and Lewis base (Scheme 10.11).

$$\text{A—B} + \text{:M}^x\text{L}_y \longrightarrow \text{A—M}^{(x+2)}\text{L}_y(\text{—B}) \quad \textit{via} \quad \text{A—B} \curvearrowleft \text{:M}^x\text{L}_y$$

Scheme 10.11

In general, the process is easy for coordinatively unsaturated metal species, in particular 16-electron (d^8 and d^{10}) metals [Ni(0), Pd(0)], since these can attain a stable 18-electron configuration as a consequence. Additionally, two metal oxidation states must be sufficiently stable. For example, tetrakis(triphenylphosphine)palladium inserts readily at 80 °C into the C–Br bond of bromobenzene to give the organometallic complex $PhPd(PPh_3)_4Br$.

In the context of organic synthesis, the process is mainly confined to the insertion of metals [mainly Pd(0) but to a lesser extent Ni(0)] into a C–X bond, where X is a leaving group such as a halide or triflate ($CF_3SO_3^-$). The order of reactivity is: I > Br ≈ CF_3SO_3 > Cl > F. Reactions of iodides and bromides can often be carried out selectively in the presence of chloride and fluoride substituents.

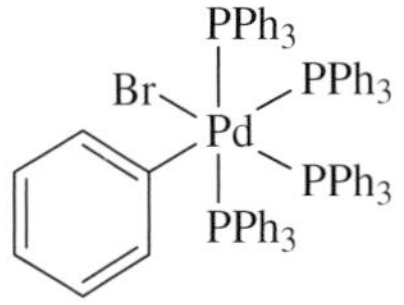

$PhPd(PPh_3)_4Br$

Palladium complexes are often written with "bystander ligands" (here the PPh_3 groups) omitted for clarity. Thus the structure is abbreviated to PhPdBr. Although these complexes, ArPdX, bear an apparent similarity to Grignard reagents, ArMgX, their reactivity is quite different. They do not undergo nucleophilic addition to carbonyl groups and are often stable to aqueous media and protic solvents.

Trifluoromethanesulfonic acid, CF_3SO_3H, is a stronger acid than H_2SO_4. The trifluoromethanesulfonate anion, $CF_3SO_3^-$, usually called "triflate", OTf, is therefore a very weak base and an excellent leaving group.

10.3.2 Reductive Elimination

Reductive elimination is the converse of oxidative addition and involves the elimination of a molecule A–B from a complex in which the groups A and B are separately bonded to the metal in a *cis* relationship (Scheme 10.12). The process involves a decrease in metal oxidation state and coordination number by two units.

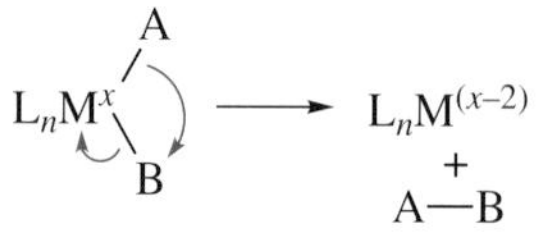

$$L_nM^x(A)(B) \longrightarrow L_nM^{(x-2)} + A{-}B$$

Scheme 10.12

10.3.3 β-Elimination

In general, transition metal organometallic species that contain an sp^3 carbon bearing a hydrogen atom β to the metal will rapidly eliminate the β-hydrogen to form an alkene and a metal hydride (Scheme 10.13).

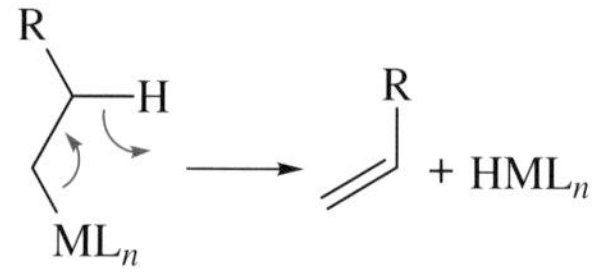

$$RCH(H)CH_2ML_n \longrightarrow RCH{=}CH_2 + HML_n$$

Scheme 10.13

Although the process, which is called β-hydride elimination or simply β-elimination, is useful in organic synthesis, it competes with other reactions, thereby limiting its value. In general, since β-elimination is rapid, transition metal mediated reactions of species bearing β-hydrogens often fail.

10.3.4 Transmetallation

When treated with a second metal complex, certain organometallic species undergo a process in which the organic ligand is transferred from one metal to the other. This is known as transmetallation (Scheme 10.14).

The mechanism by which this occurs is not fully understood, but it may be metal and ligand dependent. A simple representation is as a concerted process.

$$R{-}ML_x + X{-}M'L'_y \longrightarrow R{-}M'L'_y + XML_x$$

Scheme 10.14

A concerted reaction is one which occurs in a single step, without any intermediates. The electron movements take place simultaneously:

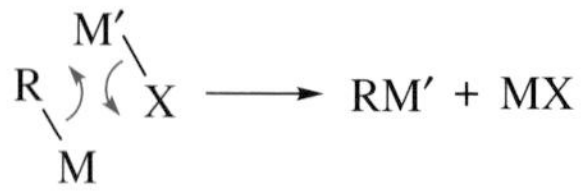

$$R(M)\cdots M'{-}X \longrightarrow RM' + MX$$

10.3.5 Insertion Reactions

Some transition metal organometallic compounds undergo addition reactions to carbon–carbon multiple bonds. The process involves a *syn*

1,2-insertion and proceeds *via* the intermediacy of a π-complex. The complex is best represented as a hybrid of the π-complex and the metallacyclopropane (Scheme 10.15). Its true nature lies somewhere between the two extreme forms.

In *syn* addition, the two new bonds, here to the metal, are formed on the same face of the double bond. This reaction is similar to the addition of dichlorocarbene, $:CCl_2$, to an alkene and to the epoxidation of alkenes.

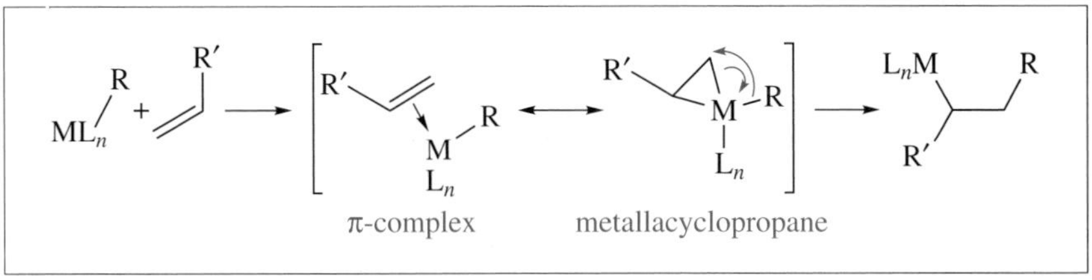

π-complex metallacyclopropane

Scheme 10.15

Complexation to a metal activates a double bond towards the addition of a nucleophilic species (Scheme 10.16). The metal has modified the behaviour of the alkene, which would normally undergo addition reactions with electrophiles.

This is comparable to epoxide ring opening:

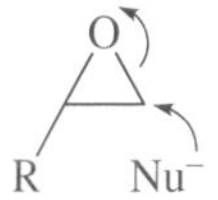

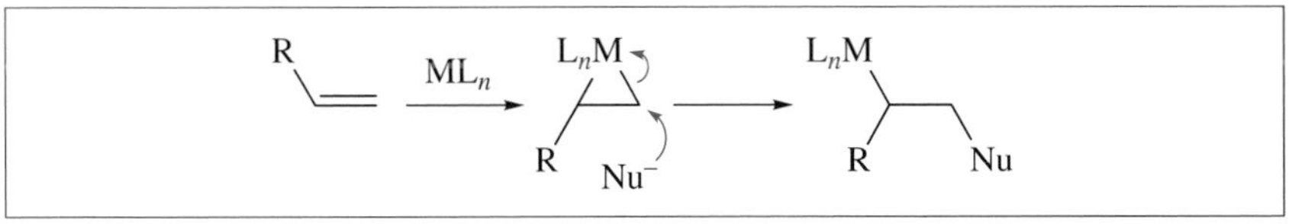

Scheme 10.16

10.4 Aryl Coupling Reactions

10.4.1 The Ullmann Coupling

Symmetrical biaryls can be formed by the coupling of two molecules of an aryl halide in the presence of copper metal (Scheme 10.17).

The notation

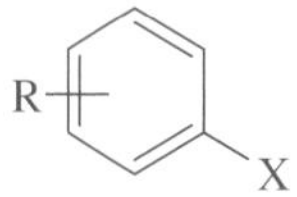

indicates that the substituent R does not occupy a specific position in the ring.

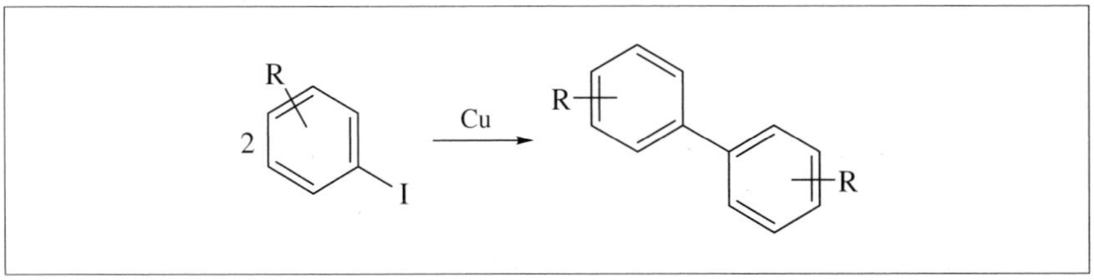

Scheme 10.17

This reaction is known as the **Ullmann coupling**. It is believed to involve the intermediacy of aryl copper complexes rather than radical species. The reaction is best suited to the preparation of symmetrical biaryls ("homo-coupled" products). Attempts to couple two different halides (Ar^1X and Ar^2X) in this way can lead to mixtures of the desired cross-coupled product (Ar^1–Ar^2) and the two homo-coupled species (Ar^1–Ar^1 and Ar^2–Ar^2).

"Homo-coupled" means that two identical species have joined together. "Cross-coupling" involves two different species.

10.4.2 The Stille and Related Reactions

The synthesis of unsymmetrical biaryls **8** from two monoaryl species involves the coupling of a metallated aromatic molecule **6** with an aryl halide or triflate **4** under the action of palladium(0) catalysis. The reaction involves a catalytic cycle in which palladium(0) inserts into the C–halogen bond *via* an oxidative addition to generate an arylpalladium(II) species **5** (Scheme 10.18). This undergoes a transmetallation with the metallated component, producing a biarylpalladium(II) complex **7**. The biaryl product is formed by reductive elimination. In the process, Pd(0) is regenerated and this can then react with a second molecule of aryl halide. Pd(0) is therefore a catalyst for the reaction.

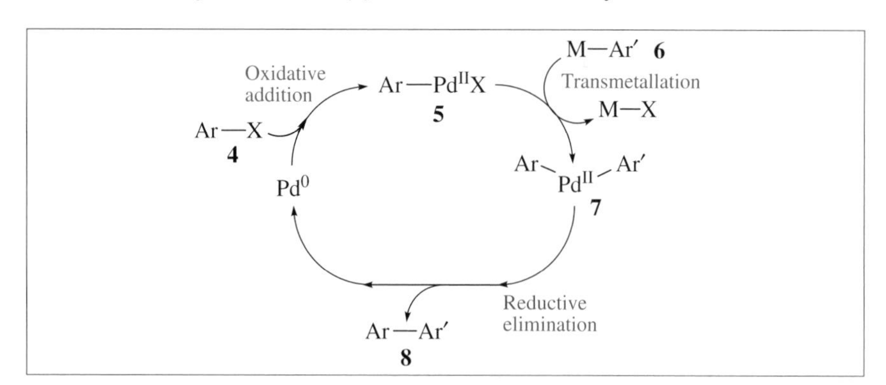

Scheme 10.18

The stannane coupling partner in these reactions can be prepared by a number of methods. Particularly useful is the transmetallation from a Grignard or aryllithium compound using a trialkylchlorostannane.

$$ArSnR_3 + Ar'X \xrightarrow{Pd^0} Ar{-}Ar'$$

Scheme 10.19

The boron compounds used in Suzuki coupling reactions may be prepared by reaction of the appropriate aryllithium compound with trimethyl borate, $(MeO)_3B$, followed by reaction with hydrochloric acid.

Coupling reactions such as this can all be explained in a general way by the sequence: (i) oxidative addition; (ii) transmetallation; (iii) reductive elimination.

There are a number of commonly used reactions of this type and many are named after the chemists who discovered them. They differ, among other things, in the exact nature of the metallic component. One of the first to be developed extensively was the **Stille coupling**, which specifically involves the coupling of an arylstannane with an aryl halide or triflate under the action of palladium catalysis (Scheme 10.19).

One drawback of the Stille coupling is that the tin by-products are toxic and are not easily removed from the product. A solution to this problem developed by Suzuki uses a boronic acid in place of the organotin compound. The boron-containing by-products are innocuous and are easily removed because of their solubility in water. The **Suzuki coupling** has found widespread use in organic synthesis (Scheme 10.20). These reactions are extremely important and the methodology is extensively used, particularly in the search for new pharmaceutical products.

Reactions of this type are not limited to two aryl components. Phenylethenes can be prepared in this way, with either the aromatic or the alkenic component being used as the organometallic compound

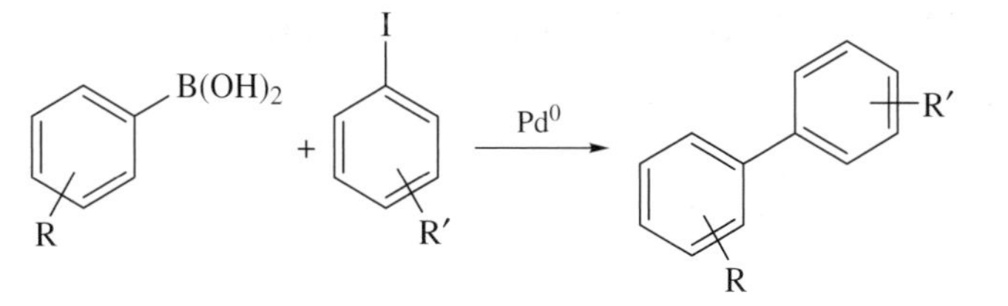

Scheme 10.20

(Scheme 10.21). It is quite common to use M to designate a metallic function. In Scheme 10.21, M represents tin and boron functions. The mechanism is analogous to that described previously.

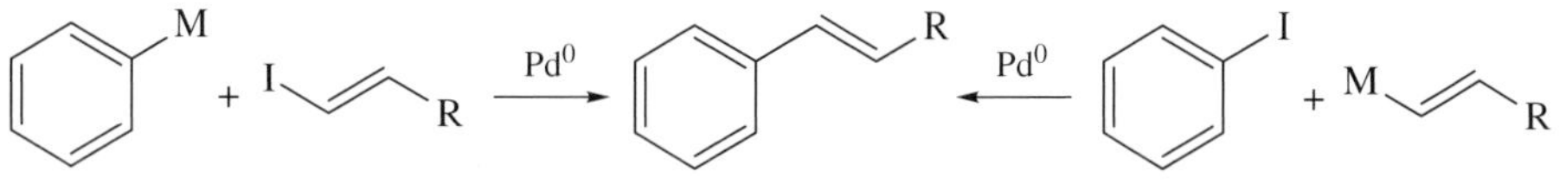

Scheme 10.21

Neither component needs to be aromatic and, although not appropriate to this book, these processes are widely applicable to C–C bond-forming reactions. However, the components should not contain β-hydrogen atoms bound to sp^3 carbons, since these can undergo β-hydride elimination.

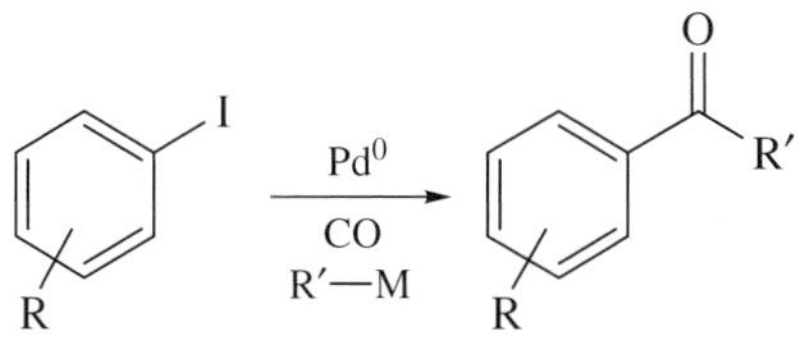

The reactions may also be carried out under an atmosphere of carbon monoxide, CO (Scheme 10.22), when the usual catalytic cycle occurs. CO inserts easily into the palladium complex Ar–Pd^{II}–X. The aryl ligand migrates on to the carbonyl group to form a metal-acyl species, X–Pd^{II}–C(O)Ar. A transmetallation–reductive elimination sequence follows, forming the ketone and regenerating the Pd^0 catalyst.

Scheme 10.22

10.4.3 The Heck Reaction

The direct coupling of an aryl halide with an alkene to produce a phenylethene is known as the **Heck reaction** (Scheme 10.23). The mechanism involves coordination of the alkene to the palladium to form a π-complex **9** with which the arene ligand can react. A variety of substituents on the alkene is compatible with the reaction.

The Heck reaction differs from Stille and Suzuki coupling in that no stoichiometric metallic coupling partner is used. Also, a β-hydride elimination rather than a reductive elimination of two groups generates the product.

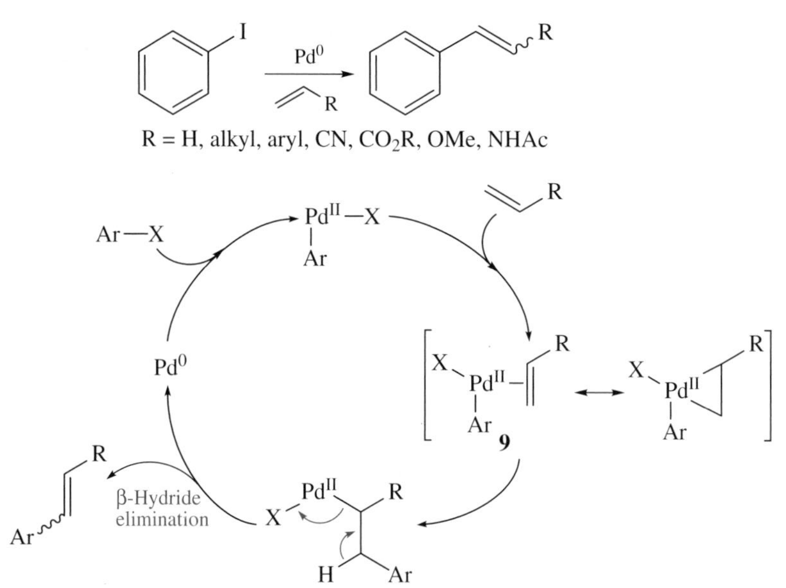

$R = H$, alkyl, aryl, CN, CO_2R, OMe, NHAc

Scheme 10.23

In the Ullmann synthesis, the presence of an amino group results in the formation of a diarylamine:

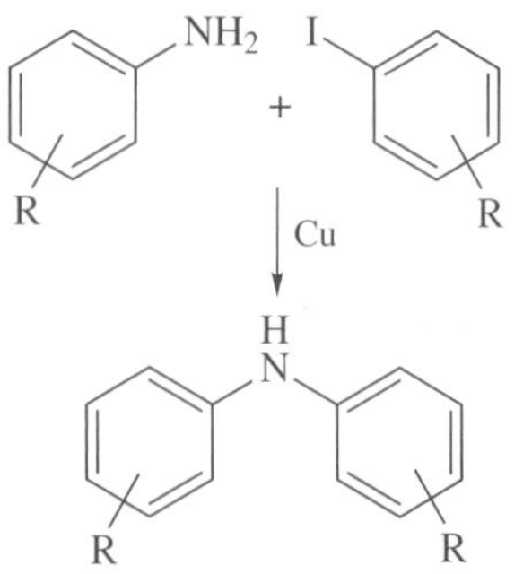

A bidentate ligand possesses two atoms which have an electron pair that can be shared with a metal. On complexation, two bonds are formed to the metal and a ring results. This process is called chelation. A commonly used example is BINAP:

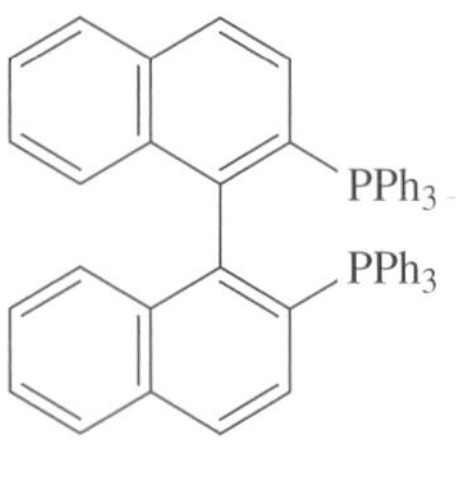

10.4.4 Amination Reactions

An aryl halide can also be coupled to an amine using metal catalysis. The reaction represents an alternative to the classical methods for the synthesis of aryl amines, such as reduction of nitro groups and nucleophilic aromatic substitution (see Chapter 8).

The reaction is particularly useful for the synthesis of biaryl amines, some of which are of value as drugs, dyes and agrochemicals, and which are often inaccessible directly by other methods.

An amine can be coupled with an aryl bromide, iodide or triflate in the presence of a palladium catalyst, a base, typically $KOBu^t$ or $CsCO_3$, and a ligand such as the bidentate phosphine BINAP. These reactions are known as **Buchwald** or **Buchwald–Hartwig reactions** (Scheme 10.24). A catalytic cycle is again involved, with the amine displacing X from $Ar–Pd^{II}–X$ to form $Ar–Pd^{II}–\overset{+}{N}HR_2$. Abstraction of a proton by the base produces $Ar–Pd^{II}–NR_2$, which undergoes a reductive elimination.

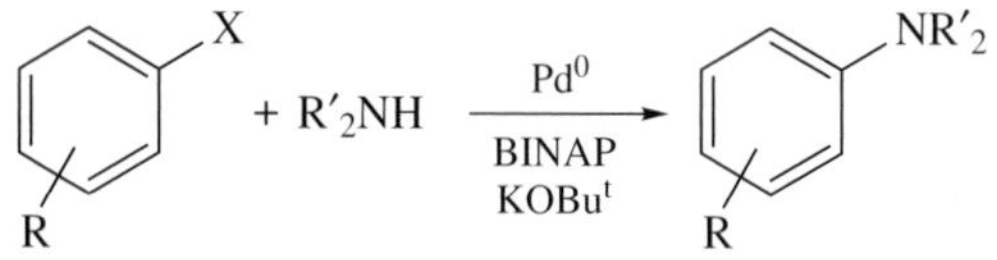

Scheme 10.24

Worked Problem 10.2

Q Aryl ethers may be prepared by coupling an aryl bromide with an alcohol in the presence of a strong base (Scheme 10.25). Suggest a catalytic cycle for the process.

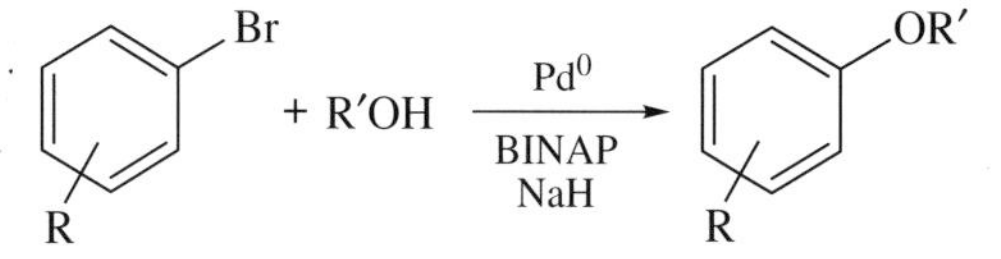

Scheme 10.25

A The mechanism involves a catalytic cycle in which the key step is displacement of the halide by the alkoxide, generated by deprotonation of the alcohol by hydride (Scheme 10.26).

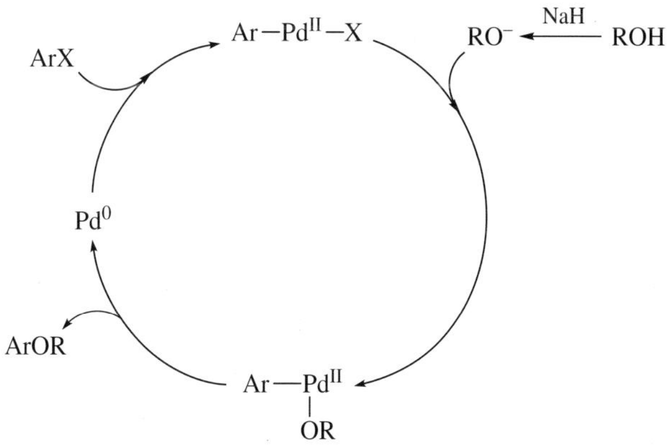

Scheme 10.26

10.5 Arene–Chromium Tricarbonyl Complexes

Arenes form η^6-complexes with a number of transition metals (*e.g.* Cr, Mo, W, Fe). Complexes of chromium have found widespread application because of their ease of synthesis, stability, easy removal of the ligands and usefulness in synthesis.

The term η^6 (eta) refers to the number of carbon atoms that are involved in bonding to the metal, in this case 6, and is called the **hapticity number**.

10.5.1 Preparation and Structure

They may be prepared by heating $Cr(CO)_6$ or $Cr(CO)_3(NH_3)_3$ in the arene as solvent (Scheme 10.27) or, when use of excess arene is undesirable, by exchange with the naphthalene complex **10**. The procedure works well for electron-rich arenes, but is of no value for electron-deficient aromatic compounds. Decomplexation can subsequently be

In the complex, the metal is bound to all six atoms of the aromatic ring:

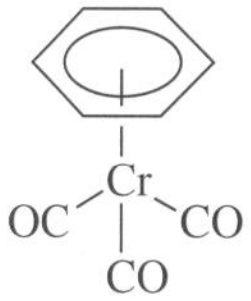

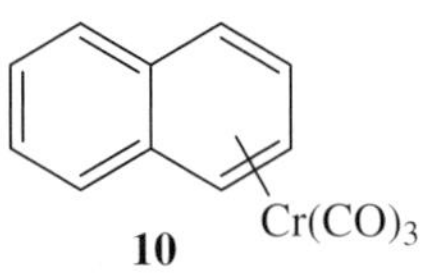

$Cr(CO)_3$
10

achieved by treatment with mild oxidants such as I_2, Fe^{III}, Ce^{III} or even by air oxidation.

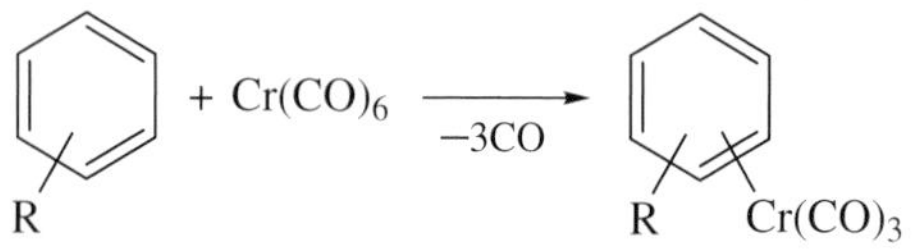

+ $Cr(CO)_6$ → −3CO

R R $Cr(CO)_3$

Scheme 10.27

Complexation has a marked effect on reactivity because it removes electron density from the π-cloud. This makes both the ring and the side chain of the arene acidic and therefore susceptible to nucleophilic attack. The electron-withdrawing effect of the chromium is comparable to that of a nitro group (see Chapter 7).

10.5.2 Reaction with Organolithium Reagents

Treatment of the Cr complex with MeLi or BuLi usually leads to deprotonation as a consequence of the acidity of the ring. The lithiated species reacts with electrophiles in the usual manner (see Section 10.1.3) (Scheme 10.28). This route is another means of deprotonation of aromatic rings and may sometimes be more convenient than directed metallation or halogen–metal exchange, especially if the precursors to such species are not readily accessible.

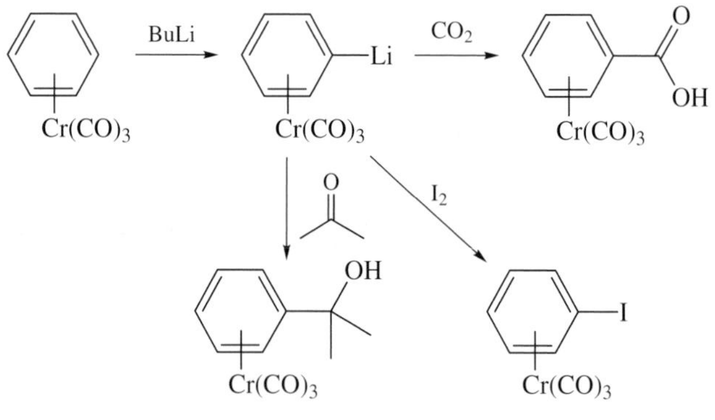

BuLi → Li → CO_2 → O OH

$Cr(CO)_3$ $Cr(CO)_3$ $Cr(CO)_3$

O I_2

OH I

$Cr(CO)_3$ $Cr(CO)_3$

Scheme 10.28

The chromium substituent activates the halogen to nucleophilic displacement in much the same way as does a nitro group (see Chapter 7).

Complexed aryl halides undergo ready displacement of the halide by nucleophiles such as alkoxides, amines and stabilized carbanions to form substituted benzenes (Scheme 10.29).

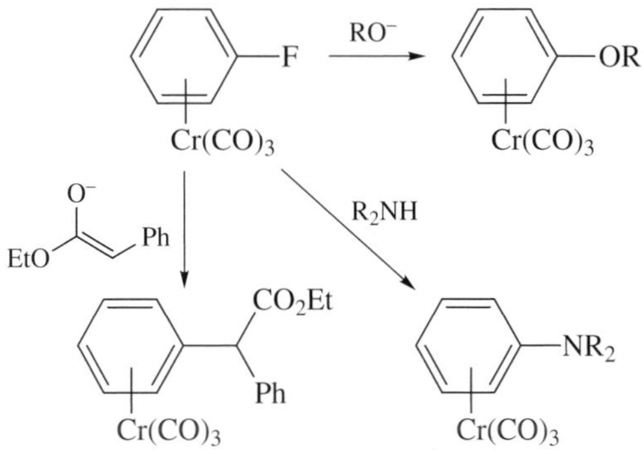

Scheme 10.29

10.5.3 The Dötz Reaction

A further application of chromium complexes in aromatic chemistry allows the construction of a new aromatic ring. In the **Dötz benzannulation**, an alkyne adds to an unsaturated alkoxychromium carbene **11** to give a hydroquinone–chromium complex **12**. Decomplexation yields the aromatic compound (Scheme 10.30).

The chromium carbene may be prepared by addition of a vinyllithium to chromium hexacarbonyl followed by alkylation with a reactive electrophile such as trimethyloxonium tetrafluoroborate (Meerwein's salt, $Me_3O^+BF_4^-$).

Carbenes are formally divalent carbon species and as such are very reactive. Certain transition metals stabilize the carbene. The metal–carbene complexes which result, some of which are stable, are called "carbenoids", since an actual carbene is no longer present.

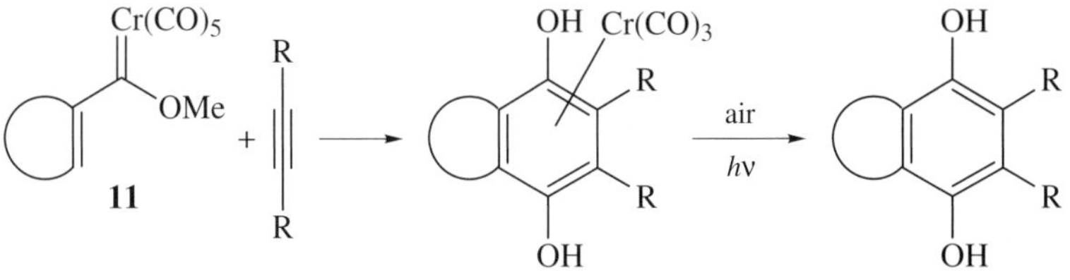

Scheme 10.30

The annulation reaction is formally a [3 + 2 + 1] cycloaddition of the carbene, alkyne and a CO molecule. The connectivity is shown in **13**.

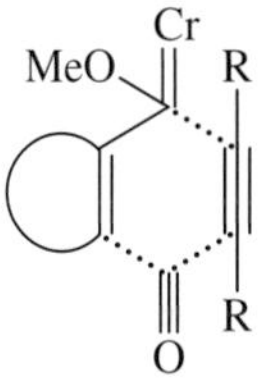

13

Summary of Key Points

1. Aryl halides react with Mg and Li to form organometallic compounds that contain electron-rich carbon centres.

2. Substituents with a pair of electrons available for coordination to Li direct metallation to the adjacent position: directed orthometallation.

3. Alcohols, aldehydes, ketones and carboxylic acids can be prepared from Grignard reagents, ArMgBr, and organolithium compounds, ArLi, by reaction with molecules with an electron-deficient site.

4. Pd(0) complexes catalyse reactions of aryl halides with various organometallic compounds which lead to biaryls, phenylethenes, ketones, amines and ethers.

5. Arene–chromium tricarbonyl complexes are electron-deficient aromatic species. They are readily deprotonated and are easily attacked by nucleophiles.

6. Chromium carbenes react with alkynes to form an aromatic ring.

Problems

10.1. Explain how bromobenzene can be converted into (a) *N*-benzyl-*N*-methylaniline using Pd catalysis and (b) propoxybenzene (phenyl propyl ether) using $Cr(CO)_3$.

10.2. Account for the transformations in (a) and (b), giving mechanistic details.

(a) 4-iodoanisole (OMe, I) + 2-methyl-3-nitrophenylboronic acid ($B(OH)_2$, NO_2) $\xrightarrow[PPh_3]{Pd^0}$ biaryl (OMe, NO_2)

(b) anisole (OMe) $\xrightarrow[2.\ I_2]{1.\ Bu^sLi}$ 2-iodoanisole (OMe, I) $\xrightarrow[\text{CH}_2=\text{CHCO}_2\text{Me}]{Pd^0,\ PPh_3}$ (OMe, CO_2Me)

10.3. Starting from iodobenzene, describe a synthetic route to each of the following compounds: (a) 2-phenylbutan-2-ol; (b) 4-methylbenzophenone; (c) 3-nitrobenzoic acid; (d) 1-phenylbutan-1-ol.

10.4. Suggest routes by which the following conversions could be achieved, using a Grignard reagent in one step of the sequence: (a) nitrobenzene → benzoic acid; (b) phenol → 4-methoxy-3-nitrobenzoic acid; (c) toluene → (4-tolyl)diphenylmethanol; (d) benzoic acid → 1-phenylbutan-1-one.

11

Oxidation and Reduction of Aromatic Compounds

Aims

By the end of this chapter you should understand:

- The methods for oxidation and reduction of the benzene ring
- Mechanisms by which these processes occur

11.1 Introduction

The unusual stability of the aromatic sextet suggests that benzene will be resistant to oxidation and reduction of the ring, since both processes will destroy the aromaticity. Although this is generally the case, both types of reaction are possible under certain conditions. This chapter is restricted to benzene and its derivatives, but other aromatic systems are more easily oxidized and reduced (see Chapter 12).

11.2 Reduction of the Benzene Ring

11.2.1 Hydrogenation

We have seen in Chapter 1 that benzene may be hydrogenated to cyclohexane, although the associated loss of resonance energy makes this process more difficult than for simple alkenes. Moreover, because the initial product, cyclohexadiene, is reduced more rapidly than benzene, hydrogenation results in complete rather than partial reduction (Scheme 11.1).

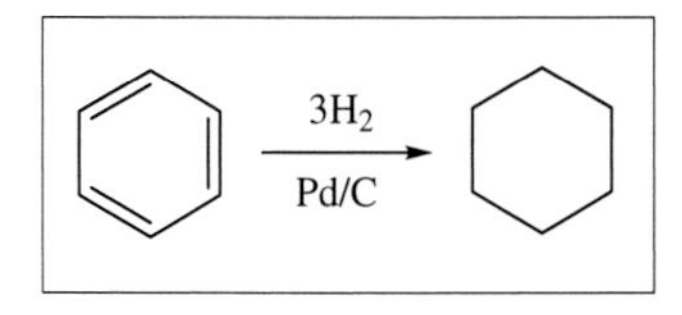

Scheme 11.1

Cyclohexane, cyclohexene and cyclohexadiene are used as hydrogen sources in the hydrogenation of alkenes to alkanes, when they are themselves oxidized to benzene. In these reactions, the driving force is the formation of the aromatic ring.

11.2.2 Alkali Metal–Ammonia Reduction

The partial reduction of arenes can be achieved using the Birch reduction. An alkali metal (lithium, sodium or potassium) is dissolved in liquid ammonia in the presence of the arene, an alcohol, such as 2-methylpropan-2-ol (*tert*-butyl alcohol) and a co-solvent to assist solubility.

The radical anion **1**, and the radical **2** and the derived anion, are not localized structures. The electrons are delocalized over the π-system and appropriate resonance forms can be written. Protonation occurs at the site of greatest charge density.

A solution of sodium in ammonia may be considered as a source of solvated electrons. The alcohol functions as a proton source. The aromatic molecule accepts an electron from the solution to form a radical anion **1**, protonation of which by the alcohol forms the radical **2** (Scheme 11.2). Acceptance of a second electron generates a new carbanion, which is also protonated and gives the 1,4-diene **3**. The overall transformation is reduction of the aromatic compound to the 1,4-diene.

Scheme 11.2

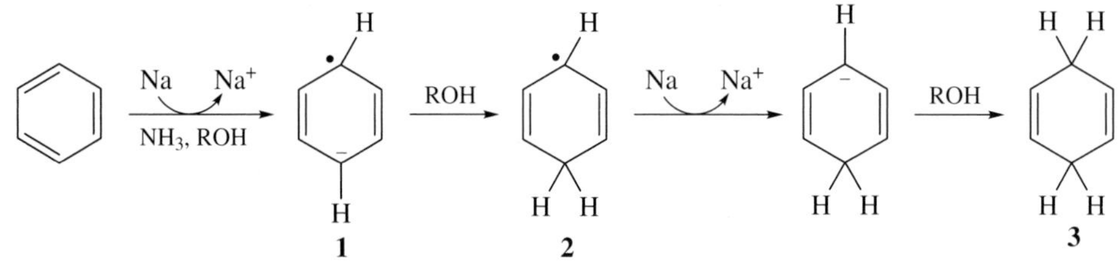

Ionization potential is the energy required to remove an electron from an atom:

$Li \xrightarrow{-e^-} Li^+$ 520 kJ mol^{-1}

$Na \xrightarrow{-e^-} Na^+$ 490 kJ mol^{-1}

$K \xrightarrow{-e^-} K^+$ 420 kJ mol^{-1}

It might be expected that the more reactive metals would be those with the lower ionization potential, but in practice lithium is the most reactive and potassium the least in the reduction of benzene. This behaviour may be a consequence of the greater solubility of lithium in ammonia.

Methoxybenzene (anisole) is reduced more rapidly than benzene under Birch conditions. This is an exception to the rule, the reasons for which are not clear.

Substituted arenes also undergo Birch reduction. In general, electron-withdrawing substituents make the arene more susceptible to reduction, while the opposite applies for electron-donating substituents. The presence of a substituent can lead to two isomeric products, in which the substituent is either of a vinylic or an allylic type. In general, electron-withdrawing groups lead to the allylic product **4** and electron-donating substituents give the vinylic product **5** (Scheme 11.3).

A vinyl group is CH_2=CH– and so a vinylic product is CH_2=CH–X. An allylic group is CH_2=CH–CH_2–; hence an allylic product is CH_2=CH–CH_2–X.

This observed selectivity is attributed to the relative stabilities of the intermediate radical anions. With an electron-withdrawing group present, the anion is most stable when localized on the carbon bearing the substituents, because of the additional stabilization by the substituent. This is illustrated in Scheme 11.4 for an ester substituent. The anion is stabilized by delocalization in an analogous manner to an enolate (see Chapter 6).

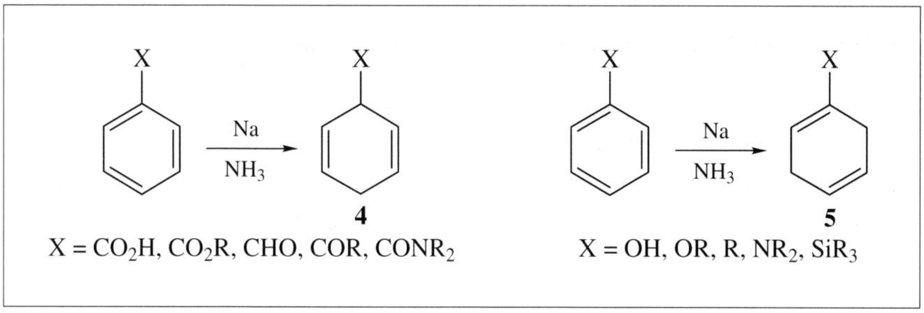

$X = CO_2H, CO_2R, CHO, COR, CONR_2$ — Na, NH_3 → **4**

$X = OH, OR, R, NR_2, SiR_3$ — Na, NH_3 → **5**

Scheme 11.3

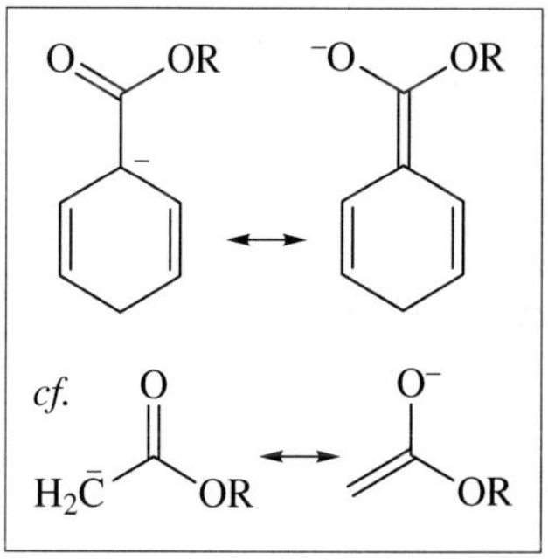

In contrast, an electron-donating group (EDG) would destabilize such an anion and so the negative charge is localized at C-2 in order to minimize the high-energy interaction between two adjacent electron-rich sites. Thus **6** is more stable than **7**.

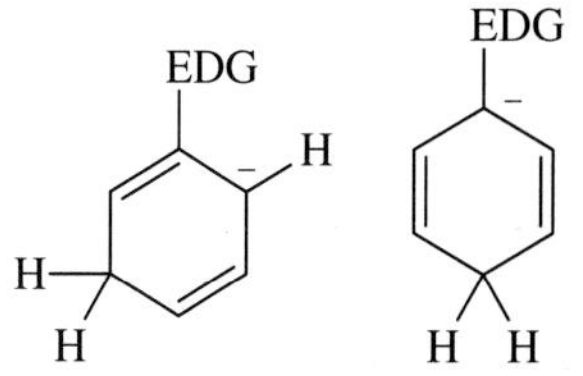

The intermediate anions may be trapped by other electrophiles besides a proton. For instance, treatment of methyl benzoate with sodium–ammonia gives the enolate **8**, which is alkylated by iodomethane to give the reduced compound **9** (Scheme 11.5). The methyl group is located α to the electron-withdrawing ester group as predicted above.

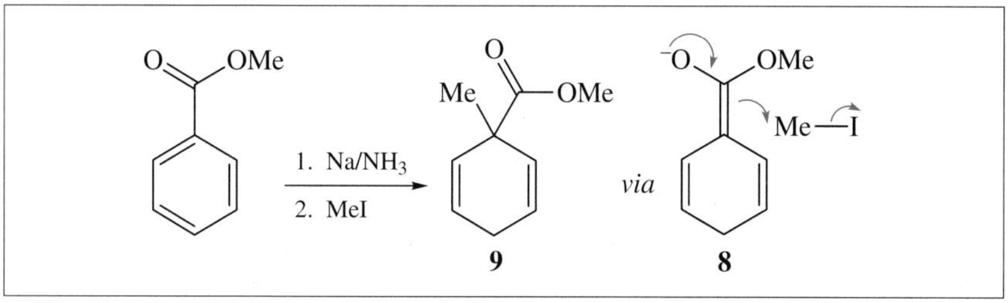

cf.

Scheme 11.4

EDG, H, **6**; EDG, H H, **7**

1. Na/NH_3 2. MeI → **9** *via* **8**

Scheme 11.5

Benzyl ethers are cleaved under both hydrogenation and Birch reduction conditions. Both methods are used in the deprotection of benzyl-protected alcohols. The by-product is toluene:

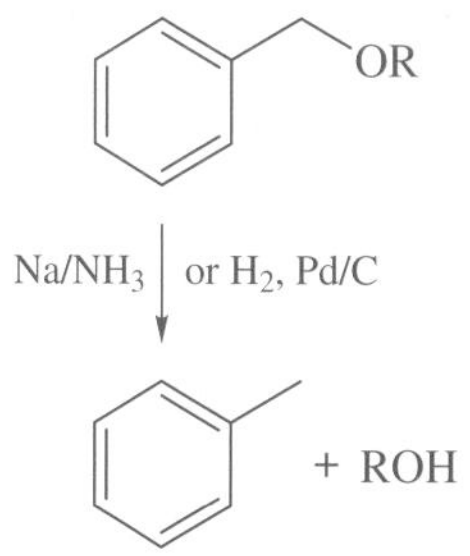

Na/NH_3 or H_2, Pd/C → + ROH

11.3 Oxidation of the Benzene Ring

11.3.1 Quinones

Except under extreme conditions, oxidation of the benzene ring requires the presence of strongly electron-donating groups such as hydroxyl or amino. These groups are simultaneously oxidized. The best known products of this oxidation process are the quinones **benzo-1,4-quinone** (*p*-benzoquinone, cyclohexadiene-1,4-dione, **10**) and **benzo-1,2-quinone** (*o*-benzoquinone, cyclohexadiene-1,2-dione, **11**).

In both of these molecules the aromaticity has been lost. However, regeneration of the aromatic ring is readily achieved. The easy reduction of benzoquinone makes it useful as an oxidizing agent. Restoration of aromaticity is the driving force in the reaction. Such quinone–hydroquinone redox systems play an important role in biological systems, besides having significant commercial value. The substituted derivatives

Vitamin K, which plays a role in the coagulation of blood, is a naphtho-1,4-quinone derivative. Co-enzymes Q, the ubiquinones, which are present in the cells of organisms, are involved in the transportation of electrons, for example in the oxidation of NADH to NAD^+. Commercially important quinones include the anthraquinone dyes (see Chapter 12) and the dyes which produce the colours in conventional photography.

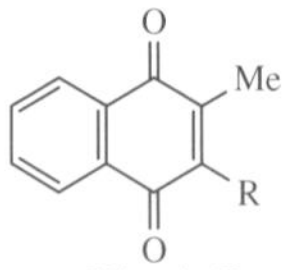

Vitamin K

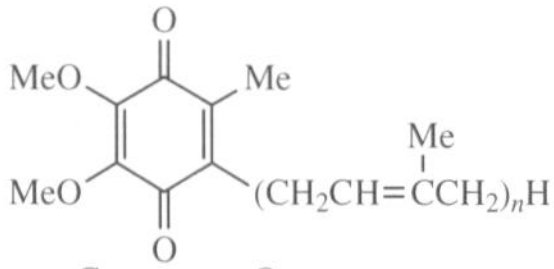

Coenzymes Q

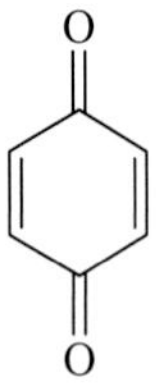

10 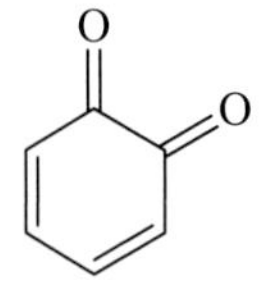 11 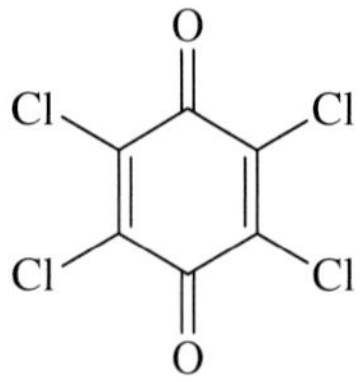 12 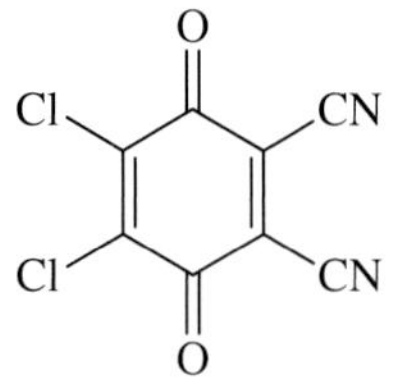 13

tetrachlorobenzo-1,4-quinone (chloranil, **12**) and 2,3-dichloro-5,6-dicyanobenzo-1,4-quinone (DDQ, **13**) are stronger oxidizing agents which are frequently used in synthesis.

Quinones can be prepared by the oxidation of phenols, dihydroxybenzenes, dimethoxybenzenes and anilines. For example,1,4-dihydroxybenzene (hydroquinone) can be oxidized in good yield using sodium chlorate in dilute sulfuric acid in the presence of vanadium pentoxide and also by manganese dioxide and sulfuric acid and by chromic acid. Other reagents which convert hydroquinones to quinones include Fremy's salt [potassium nitrosodisulfonate, $(KSO_3)_2NO$] and cerium(IV) ammonium nitrate [CAN, $Ce(NH_4)_2(NO_3)_6$].

MnO_2, H_2SO_4; Fe, H_2O; $NaClO_3$, H_2SO_4, V_2O_5

Scheme 11.6

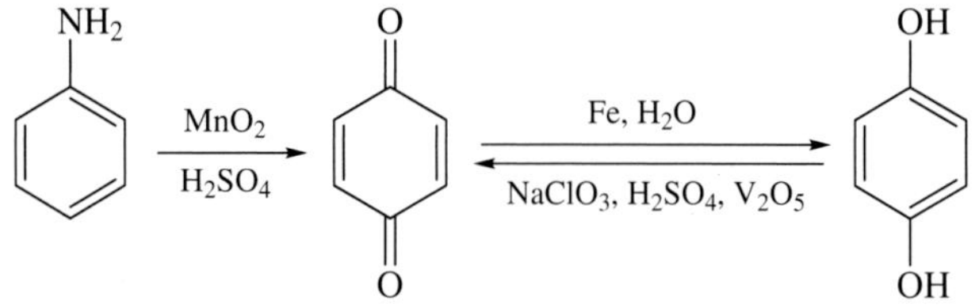

Benzoquinones are highly conjugated molecules and are consequently coloured. Benzo-1,4-quinone itself is yellow and benzo-1,2-quinones are often red. Chloranil forms glistening yellow platelets on crystallization from fuming nitric acid! Quinones occur naturally and some natural pigments contain a quinone unit

Under extreme conditions, the benzene ring can be oxidized to other products. Maleic anhydride is prepared industrially by passing a mixture of benzene and air over a vanadium pentoxide catalyst at 420 °C.

The commercial route to hydroquinone from aniline (see Chapter 4) proceeds *via* the isolated intermediate benzo-1,4-quinone (Scheme 11.6).

Benzo-1,2-quinone (**11**) is prepared by the oxidation of 1,2-dihydroxybenzene (catechol) with silver(I) oxide in diethyl ether. This compound is not very stable and is also an oxidizing agent.

11.3.2 Microbial Oxidation

Microbial oxidation of arenes is a feasible process. An example is the conversion of benzene to cylohexa-3,5-diene-1,2-diol (**14**, R = H) by the bacterium *Pseudomonas putida* (*P. putida*). The process is stereoselective and with substituted benzenes (**14**, R ≠ H) a single enantiomer is produced (Scheme 11.7). Such compounds are useful starting materials for natural product synthesis.

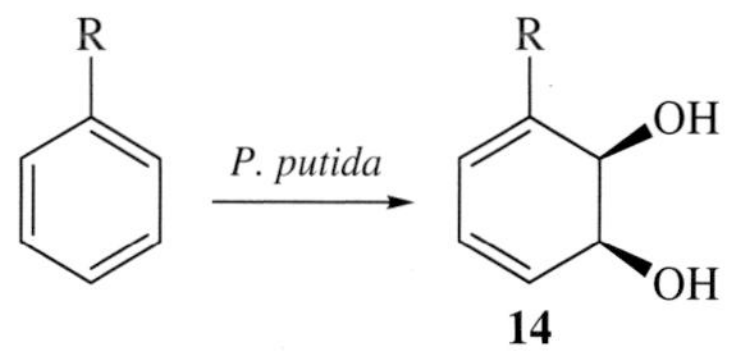

Scheme 11.7

Worked Problem 11.1

Q Explain why the products of *P. putida* oxidations are unstable.

A The cyclohexadienediols can readily regain aromaticity by eliminating water and forming the phenol (Scheme 11.8). Note that the two groups lost in this 1,2-elimination process are *anti* to each other.

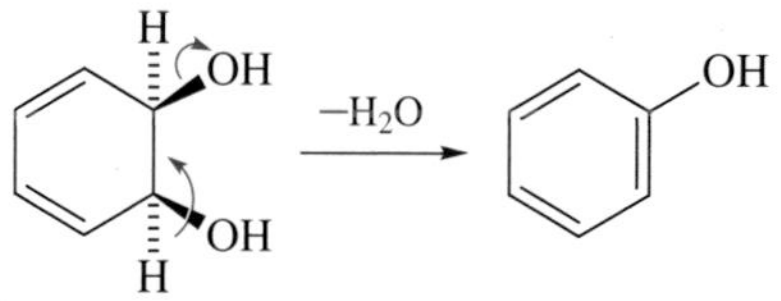

Scheme 11.8

Summary of Key Points

1. Oxidation and reduction of the stable benzene ring is difficult.

2. The Birch reduction is a useful synthetic route to dihydrobenzene derivatives.
- π-Electron-deficient benzene derivatives yield the allylic product.
- π-Electron-rich benzene derivatives give the vinylic product.

3 Benzoquinones, derived by oxidation of phenols and amines, are efficient oxidizing agents.

4. Bacterial oxidation of benzene and its derivatives forms cyclohexadienediols in a stereoselective manner.

Problems

11.1. Predict the products of reactions (a)–(c) and account for their formation:

(a) OMe-substituted benzene $\xrightarrow[\text{Bu}^t\text{OH}]{\text{Na/NH}_3}$

(b) $CONMe_2$-substituted benzene $\xrightarrow[\text{2. PhCH}_2\text{Br}]{\text{1. Na/NH}_3}$

(c) Br-substituted benzene $\xrightarrow{\textit{P. putida}}$

11.2 Identify the intermediate and final products in reaction sequences (a)–(c):

(a) hydroquinone (OH, OH) $\xrightarrow{2Br_2}$ **A** $\xrightarrow[H_2O]{FeCl_3}$ **B** $\xrightarrow[\text{heat}]{\text{cyclohexa-1,3-diene}}$ **C**

(b) hydroquinone (OH, OH) $\xrightarrow[H_2SO_4]{Na_2Cr_2O_7}$ **F** $\xrightarrow{HCl}$ **G**

(c) phenanthrene-9,10-dione (O, O) $\xrightarrow{\text{o-phenylenediamine } (NH_2, NH_2)}$ **M**

12

Polycyclic Aromatic Hydrocarbons

Aims

By the end of this chapter you should understand:

- The methods of synthesis of the main fused aromatic systems
- Electrophilic attack on naphthalene, anthracene and phenanthrene
- The oxidation of these polycycles

12.1 Introduction

The term polycyclic aromatic hydrocarbon is usually applied to compounds where the rings are fused together as in the case of **naphthalene** (**1**), **anthracene** (**2**) and **phenanthrene** (**3**). This chapter will concentrate on the synthesis and reactions of these molecules.

Compounds such as the biphenyls and diphenylmethane are sometimes referred to as polycyclic systems, but a brief description of the chemistry of these compounds has already been given in Chapter 3.

1

2

3

There are many other fused systems, of which pyrene is a more complex example. An interesting system is the C_{60} carbon allotrope, buckminsterfullerene, which consists of 12 five- and 20 six-membered rings fused together to form the so-called "buckyball". Many polycyclic hydrocarbons are potent carcinogens.

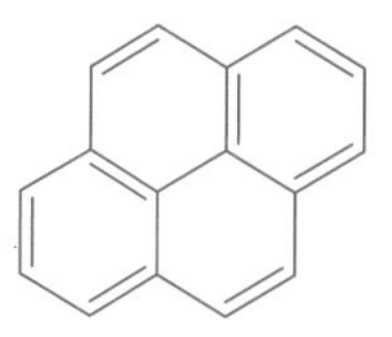

Pyrene

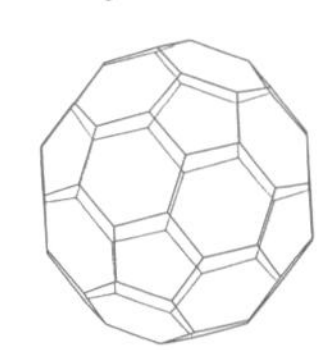

Buckminsterfullerene (C_{60})

12.2 Chemistry of Naphthalene

12.2.1 Introduction

Naphthalene (**1**) is the largest single component of coal tar at 9% and this still remains a source, although it is also produced from petroleum fractions at high temperature. Not all positions on the naphthalene ring are equivalent and the numbering of the ring is as shown in structure **1**. The positions 1 and 2 are also called the α- and β-positions.

Note that positions 1, 4, 5 and 8 are identical and all are α-positions.

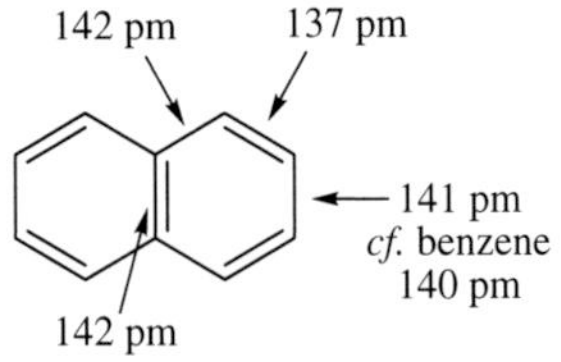

142 pm 137 pm
141 pm *cf.* benzene 140 pm
142 pm
4

It is incorrect to show the aromatic π-system in naphthalene with a circle in both rings. This representation of the π-cloud applies only to six π-electrons. Although one ring of naphthalene has six π-electrons, the other has only four π-electrons.

In contrast to benzene, the bond lengths in naphthalene are not all equal, as illustrated in **4**. The resonance energy of naphthalene is 255 kJ mol^{-1}, which is higher than, though not twice that of, benzene (151 kJ mol^{-1}). In the canonical forms **5** and **7** that contribute to the valence bond structure for naphthalene, only one of the two rings is fully benzenoid. Naphthalene is less aromatic than benzene, which accounts for its higher reactivity towards electrophilic attack compared with benzene.

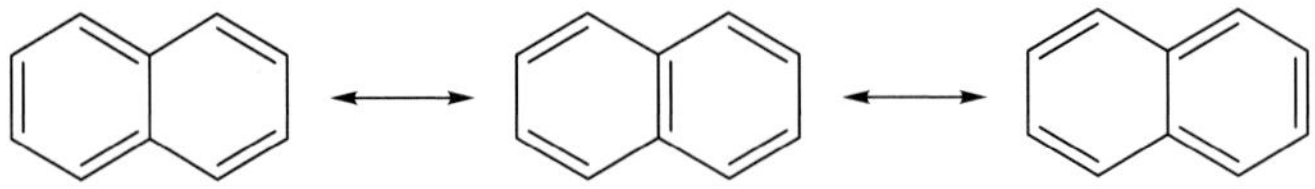

5 **6** **7**

12.2.2 Synthesis of Naphthalene

There are two main synthetic routes to naphthalene: the Haworth synthesis and a Diels–Alder approach. In the **Haworth synthesis** (Scheme 12.1), benzene is reacted under Friedel–Crafts conditions with succinic anhydride (butanedioic anhydride) to produce 4-oxo-4-phenylbutanoic acid, which is reduced with either amalgamated zinc and HCl (the Clemmensen reduction) or hydrazine, ethane-1,2-diol and potassium hydroxide (the Wolff–Kischner reaction) to 4-phenylbutanoic acid. Ring closure is achieved by heating in polyphosphoric acid (PPA). The product is 1-tetralone and reduction of the carbonyl group then gives 1,2,3,4-tetrahydronaphthalene (tetralin). Aromatization is achieved by dehydrogenation over a palladium catalyst.

Conversion of the acid into the acyl chloride and subsequent Friedel–Crafts cyclization is an alternative route to the tetralone.

Finely divided palladium has the ability to adsorb large quantities of hydrogen and is used to catalyse both hydrogenation and dehydrogenation reactions.

This route to naphthalenes is versatile. Alkyl and aryl substituents can be introduced into the 1-position through reaction of 1-tetralone with a

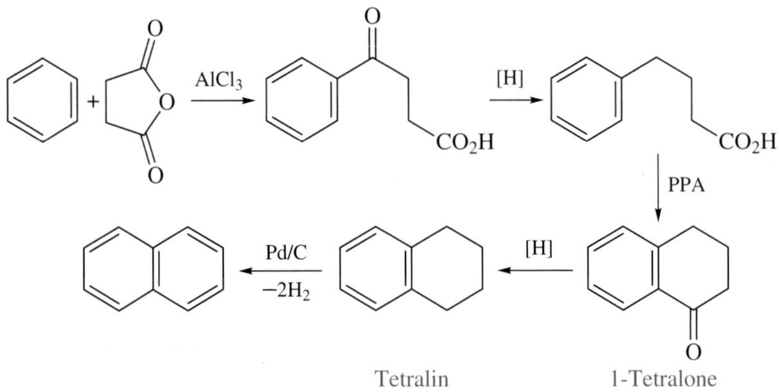

O O O
$AlCl_3$
CO_2H
[H]
CO_2H
PPA
Pd/C $-2H_2$
[H]
O
Tetralin
1-Tetralone

Scheme 12.1

Grignard reagent (see Chapter 10), followed by dehydration and aromatization (Scheme 12.2). The use of substituted benzenes in the first stage of the sequence enables variously substituted derivatives of naphthalene to be obtained. Of course, the substituents should not interfere with the Friedel–Crafts reaction with succinic anhydride.

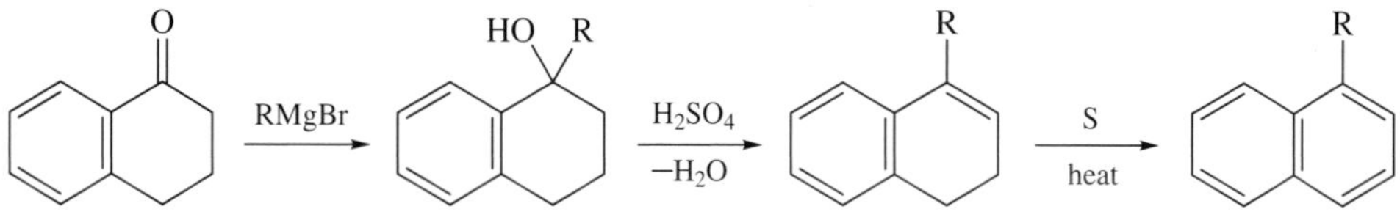

RMgBr; H_2SO_4 / $-H_2O$; S / heat

Scheme 12.2

The Diels–Alder reaction of benzo-1,4-quinone with 1,3-dienes produces adducts that may be converted to naphtho-1,4-quinones *via* enolization and oxidation (Scheme 12.3).

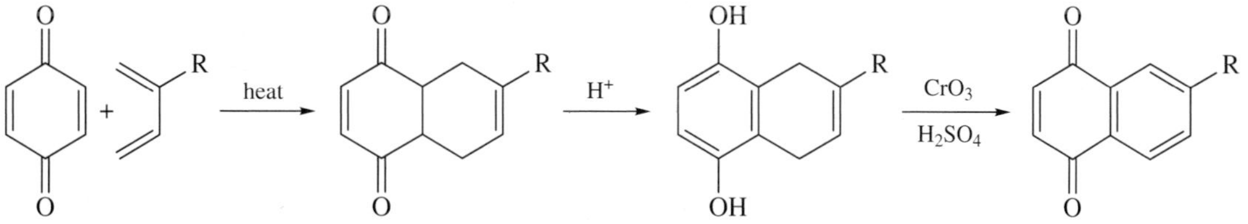

heat; H^+; CrO_3 / H_2SO_4

Scheme 12.3

12.2.3 Reactions of Naphthalene

Naphthalene is readily hydrogenated to tetrahydronaphthalene, which is used as a paint solvent, but further reduction to produce decahydronaphthalene (decalin) requires forcing conditions (Raney nickel catalyst at 200 °C). The first step in the reduction of naphthalene can be achieved by reaction with sodium in boiling ethanol, which produces 1,4-dihydronaphthalene. The tetrahydro compound is formed in the higher-boiling 3-methylbutanol (isopentyl alcohol) (Scheme 12.4).

Na, EtOH / 78 °C; Na / $Me_2CH(CH_2)_2OH$ / 132 °C

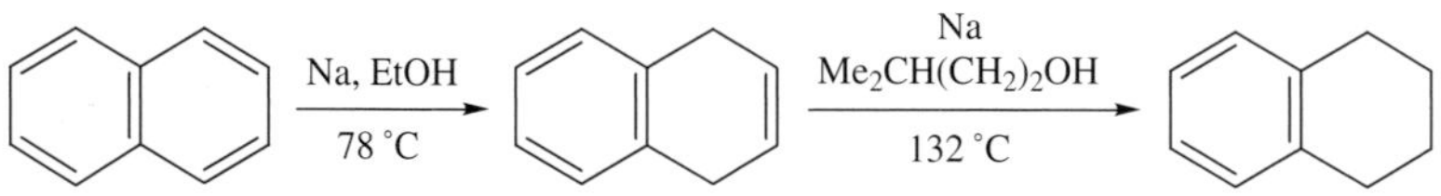

Scheme 12.4

In Chapter 2, electrophilic substitution in naphthalene was discussed, when consideration of the stability of the cationic intermediates arising

The difficulty in the reduction of tetralin is associated with the presence of the fully aromatic ring and is in keeping with the need for forcing conditions to reduce benzene to cyclohexane.

Decahydronaphthalene (decalin) can exist in *cis* and *trans* forms:

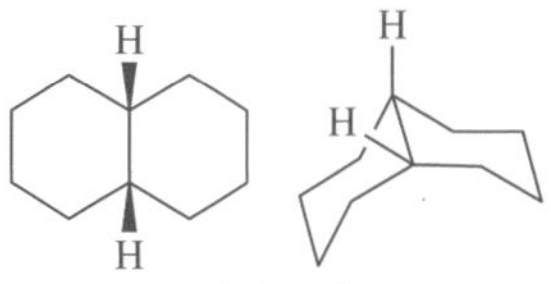

cis-decalin

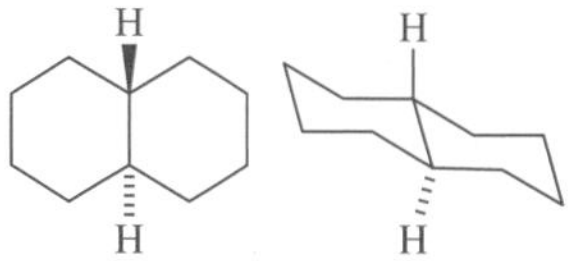

trans-decalin

The most efficient stabilization of the intermediate carbocation produced by electrophilic attack on naphthalene comes from those resonance forms which retain one fully benzenoid ring. In the case of 1-substitution, two such structures can be drawn, **8** and **9** (plus their Kekulé forms). For 2-substitution there is only one structure, **10** (and its Kekulé form). The intermediate from attack at the 1-position is the more stable and therefore the 1-substituted product is favoured.

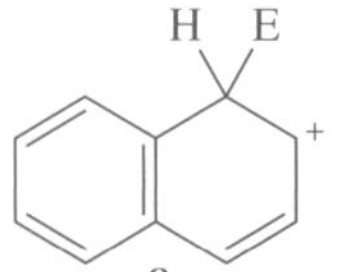

8 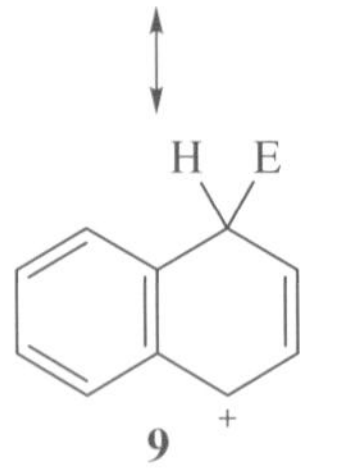 ↔ **9**

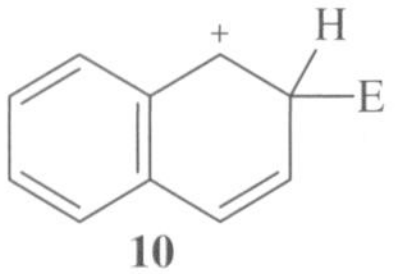

10

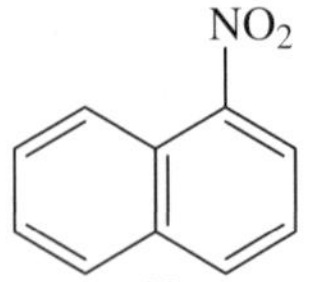

12

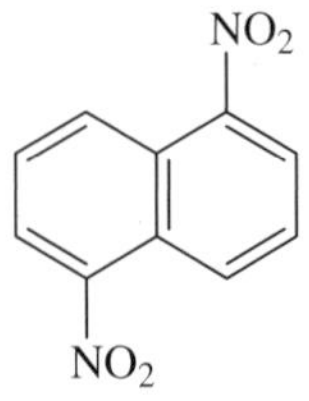

13

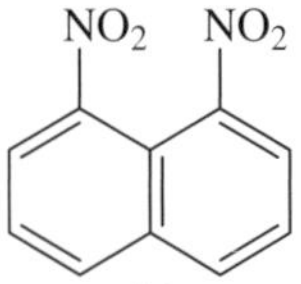

14

from attack at the 1- and 2-positions indicated that the former is favoured. Nevertheless, the energy of the two intermediates is not vastly different and under more forcing conditions attack may occur at the 2-position. Naphthalene is more reactive towards electrophiles than is benzene and hence milder conditions are generally employed.

In the case of Friedel–Crafts reactions, mild conditions are essential, since binaphthyls are formed under vigorous conditions. Reaction with acetyl chloride in tetrachloroethane in the presence of aluminium chloride gives 1-acetylnaphthalene (Scheme 12.5), although in nitrobenzene the 2-acetyl derivative **11** is the major product. Attack at the less hindered 2-position is preferred in the latter case because of the larger size of the solvated acylating species.

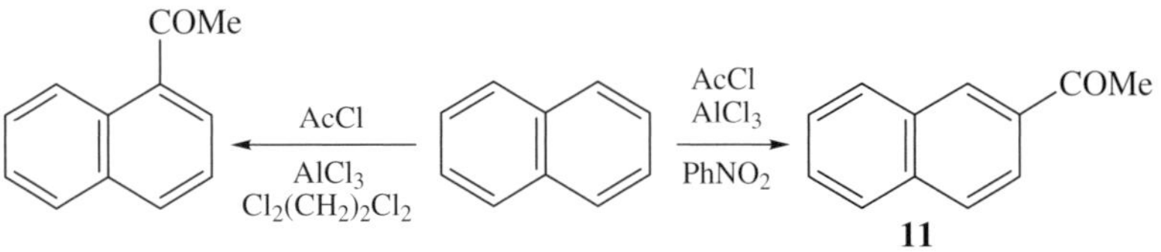

COMe ← AcCl, $AlCl_3$, $Cl_2(CH_2)_2Cl_2$ — naphthalene — AcCl, $AlCl_3$, $PhNO_2$ → COMe **11**

Scheme 12.5

Naphthalene is also easily halogenated. For example, bromination in the presence of aluminium chloride results in a 99% yield of 1-bromonaphthalene.

Nitration of naphthalene gives 1-nitronaphthalene (**12**). Further substitution does not occur in the same ring and the main products of dinitration are 1,5- (**13**) and 1,8-dinitronaphthalenes (**14**). The initial nitro group deactivates that ring to further electrophilic substitution and attack at the α-positions of the other ring therefore takes place.

The sulfonation of naphthalene by concentrated sulfuric acid at 80 °C gives naphthalene-1-sulfonic acid. At 160 °C, naphthalene-2-sulfonic acid predominates (Scheme 12.6). The 1-isomer is the more readily formed and is stable at the reaction temperature of 80 °C. However, at 160 °C, not only is naphthalene-1-sulfonic acid desulfonated, but also there is sufficient energy to convert naphthalene into the 2-sulfonic acid, which is stable even at the higher temperature. The 2-substituted isomer is more

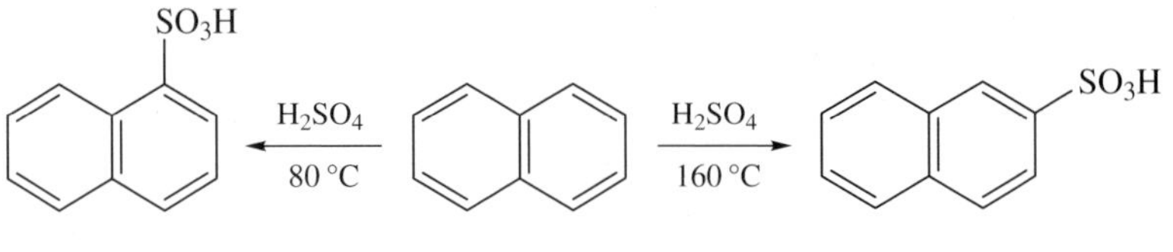

SO_3H ← H_2SO_4, 80 °C — naphthalene — H_2SO_4, 160 °C → SO_3H

Scheme 12.6

stable, probably because there is less steric hindrance between the bulky sulfonic acid group and the adjacent *ortho* H atoms (H-1 and H-3) than between the 1-sulfonic acid and the *peri* H-8 atom.

This process illustrates the concept of kinetic *versus* thermodynamic control of a reaction, with naphthalene-1-sulfonic acid being the kinetic product and the 2-sulfonic acid the thermodynamic product. The energy changes associated with these processes are illustrated in Figure 12.1.

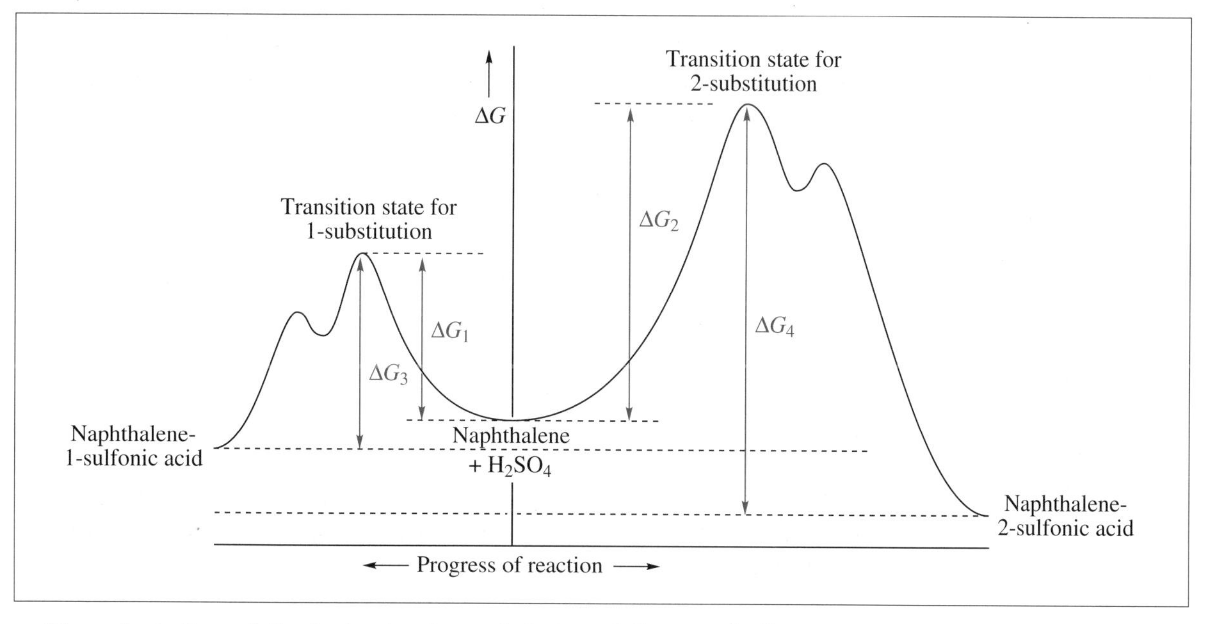

Figure 12.1 Energy profile for the sulfonation of naphthalene. ΔG_1 = energy of activation for 1-substitution; ΔG_2 = energy of activation for 2-substitution; $\Delta G_1 < \Delta G_2$, so naphthalene-1-sulfonic acid is the more easily formed and is the kinetic product. ΔG_3 = energy required to reverse formation of naphthalene-1-sulfonic acid; ΔG_4 = energy required to reverse formation of naphthalene-2-sulfonic acid; $\Delta G_4 > \Delta G_3$, so naphthalene-2-sulfonic acid is thermodynamically more stable than the 1-sulfonic acid

The orientation of disubstitution in naphthalene follows a similar pattern to that encountered in benzene:

- Electron-donating substituents at the 1-position activate the 2- and 4-positions
- Electron-donating substituents at the 2-position activate the 1-position
- Electron-withdrawing substituents direct attack to the second ring

With an electron-donating substituent at C-2, attack at C-1 is preferred since attack at C-3 would produce a Wheland intermediate in which the aromaticity in the second ring is disturbed (Figure 12.2).

Hydroxynaphthalenes, which are called **naphthols**, are readily accessible from the naphthalenesulfonic acids by heating them to fusion with solid alkali. They are also available from coal tar.

The reactions of both 1-naphthol and 2-naphthol closely resemble those of phenols. For example, both can be acylated and alkylated. 2-Naphthol is more reactive than 1-naphthol. The hydroxyl group activates the ring to electrophilic substitution. Thus, in addition to attack by the usual electrophiles, reaction with weaker electrophiles can occur,

aromaticity retained in one ring following attack at the 1-position

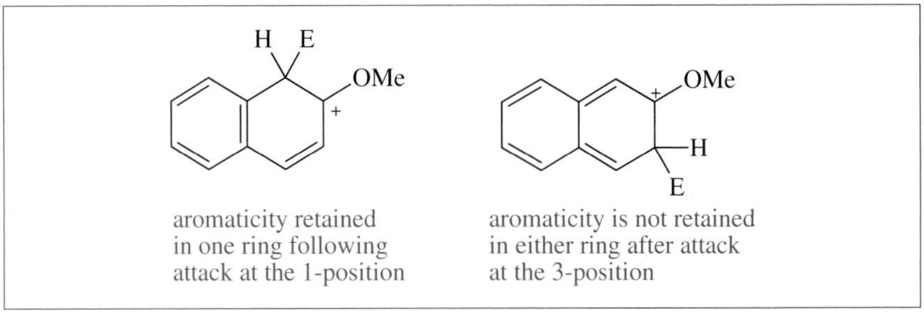

aromaticity is not retained in either ring after attack at the 3-position

Figure 12.2

as exemplified by azo coupling and nitrosation. 1-Naphthol couples with benzenediazonium chloride at the 4-position (Scheme 12.7); 2-naphthol couples at the 1-position.

In dye chemistry, the term "coupling" is used specifically to mean electrophilic attack by a diazonium salt on phenols, naphthols and anilines.

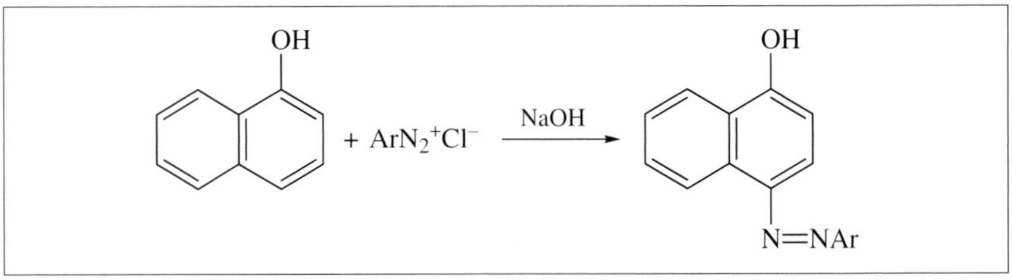

$+ ArN_2^+Cl^-$ NaOH → N=NAr

Scheme 12.7

Tautomers are isomers in which the site of a hydrogen atom and a double bond are different. The tautomers are in dynamic equilibrium with each other. We have already met keto–enol tautomerism, of which this is a variation, in Chapter 6. Imine–enamine tautomerism is another example:

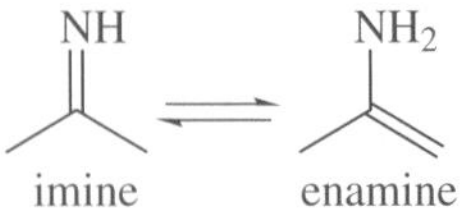

imine ⇌ enamine

Treatment of 1- and 2-naphthol with nitrous acid results in the introduction of a nitroso group at the expected positions. The products exist as a mixture of the nitroso and oxime tautomers, conjugated with the enol and keto functions, respectively (Scheme 12.8).

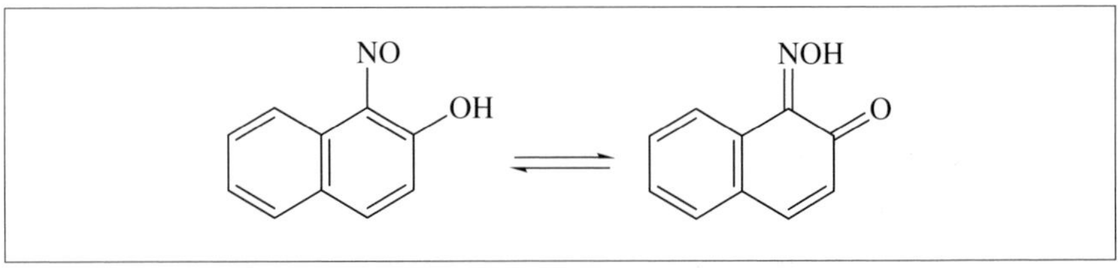

Scheme 12.8

Naphthalenesulfonic acids, naphthols and naphthylamines are important intermediates in the synthesis of commercial azo dyes. A sulfonic acid group confers water solubility on the dye.

The **naphthylamines** may be prepared by reduction of the corresponding nitro compound, but they are readily accessible from naphthols by the **Bucherer reaction**. The naphthol is heated, preferably under pressure in an autoclave, with ammonia and aqueous sodium hydrogen sulfite solution, when an addition–elimination sequence occurs. The detailed mechanism is not completely elucidated, but the Bucherer reaction is restricted to those phenols that show a tendency to tautomerize to the keto form, such as the naphthols and 1,3-dihydroxybenzene (resorcinol). Using 1-naphthol for illustration, the first step is addition of the hydrosulfite across the 3,4-double bond of either the enol or keto tautomer (Scheme 12.9). Nucleophilic attack by ammonia at the carbonyl group is followed by elimination of water. The sequence is completed by tautomerization of the imine to the naphthylamine and elimination of hydrosulfite. The Bucherer reaction is fully reversible and naphthylamines may be converted into naphthols by treatment with aqueous sodium hydrogen sulfite.

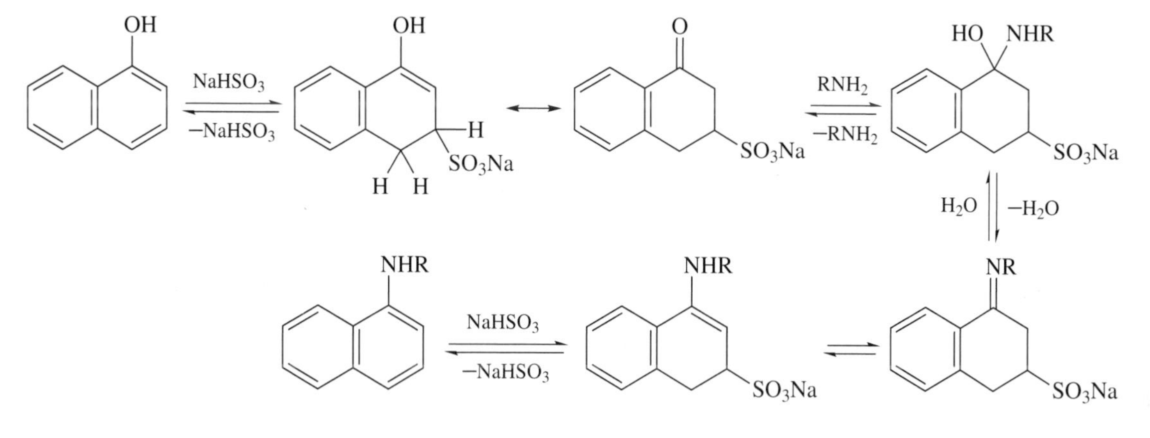

OH, $NaHSO_3$, $-NaHSO_3$, OH, H, SO_3Na, H H, O, SO_3Na, RNH_2, $-RNH_2$, HO NHR, SO_3Na, H_2O, $-H_2O$, NHR, $NaHSO_3$, $-NaHSO_3$, NHR, SO_3Na, NR, SO_3Na

Scheme 12.9

The amino group of the naphthylamines exhibits reactions typical of aniline. As with aniline, it is advantageous to acetylate the group prior to further electrophilic substitution.

Both α-naphthylamine and β-naphthylamine have been important industrial compounds with *N*-phenyl-1-naphthylamine being a rubber anti-oxidant and 2-naphthylamine being used in the azo dyestuffs industry. Unfortunately, the latter is strongly carcinogenic and the use of both naphthylamines has now been discontinued.

12.3 Chemistry of Anthracene

12.3.1 Introduction

Valence bond theory considers that anthracene is best regarded as a resonance hybrid of the four structures **15–18**. The resonance energy of anthracene is 351 kJ mol^{-1}. Examination of the canonical forms indicates that the three rings cannot all be benzenoid at the same time. It can also be seen that the central ring contains a four-carbon fragment with a relatively high degree of double bond character. The numbering system, shown in **15**, is a little unusual and was introduced during early chemical studies to indicate the special character associated with the 9- and 10-positions.

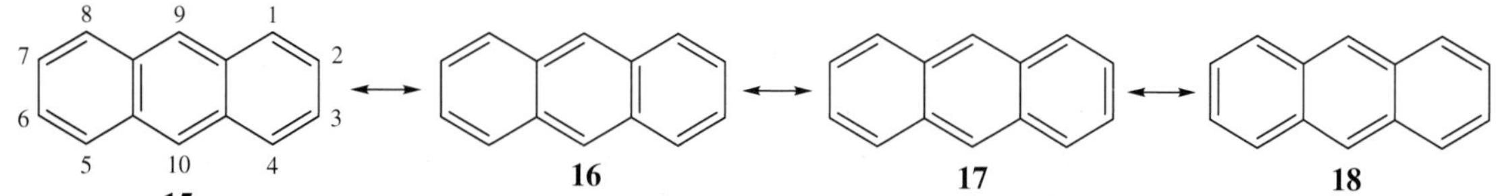

8 9 1 7 2 6 3 5 10 4

15 **16** **17** **18**

12.3.2 Synthesis of Anthracene

Although it is possible to synthesize anthracene in a number of ways using Friedel–Crafts methodology, the usual routes involve either an adaptation of the **Haworth synthesis** of naphthalene or a **Diels–Alder reaction** using naphtho-1,4-quinone as the dienophile.

Friedel–Crafts reaction of phthalic anhydride with benzene in the presence of aluminium chloride followed by cyclization under acidic conditions gives anthra-9,10-quinone (Scheme 12.10). Distillation over zinc dust gives anthracene.

Scheme 12.10

Anthra-9,10-quinone is the eventual product from the cycloaddition of buta-1,3-diene to naphtho-1,4-quinone after oxidation of the tetrahydroanthraquinone adduct.

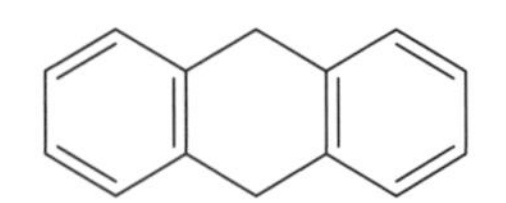

Since both the anthraquinone and the dihydroanthracene have two non-conjugated benzenoid rings, the resonance stabilization energy of each molecule is 2 × 150 kJ mol^{-1}. The resonance energy of anthracene is 351 kJ mol^{-1} and so the loss of resonance stabilization energy associated with the oxidation and reduction reactions is only of the order of 50 kJ mol^{-1}.

12.3.3 Reactions of Anthracene

Much of the chemistry of anthracene is associated with the special character of the central fragment of the system. Thus anthracene is easily oxidized to form anthra-9,10-quinone. Reduction to 9,10-dihydroanthracene is readily achieved with Na/EtOH. In both cases the product contains two fully non-conjugated benzenoid rings.

Similarly, halogenation of anthracene involves addition at the 9,10-positions, giving 9,10-dichloro-9,10-dihydroanthracene which on heating loses HCl to form 9-chloroanthracene.

Anthracene cannot be nitrated with nitric acid because of its easy oxidation to anthraquinone, although 9-nitroanthracene can be isolated from nitration in acetic anhydride at room temperature.

The dienic character of the central ring is illustrated by the reaction of anthracene with dienophiles in Diels–Alder reactions. For example, *cis*-butenedioic anhydride (maleic anhydride) reacts readily; when benzyne is generated in the presence of anthracene, triptycene (**19**) is produced.

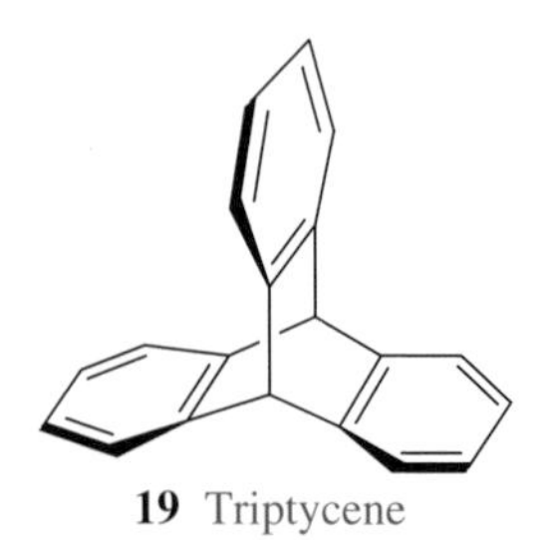

19 Triptycene

Derivatives of anthraquinone are important as dyestuffs for the coloration of a variety of fabrics.

12.4 Chemistry of Phenanthrene

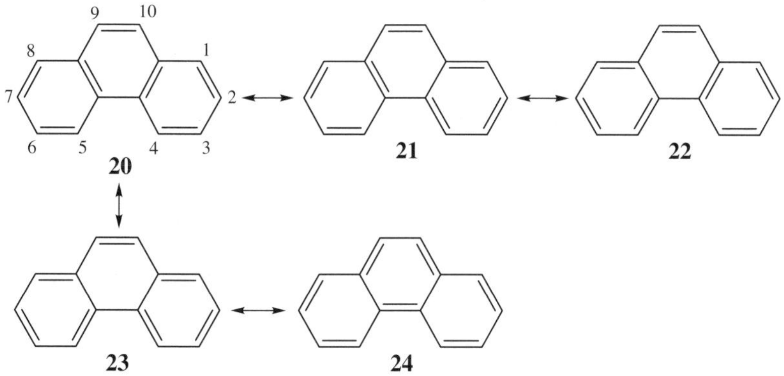

Examples of anthraquinone dyes are Disperse Blue 3 and Disperse Red 60. These dyes, which are used in the colouration of polyester fabrics, are prepared for the dyeing process as a fine dispersion in water.

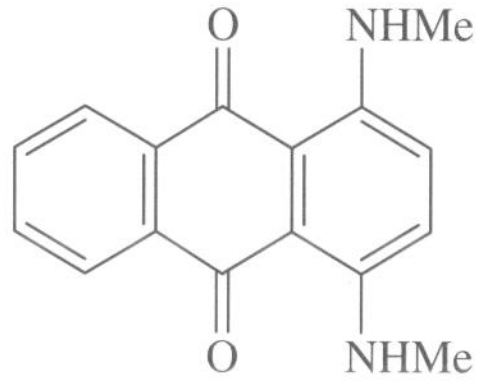

Disperse Blue 3

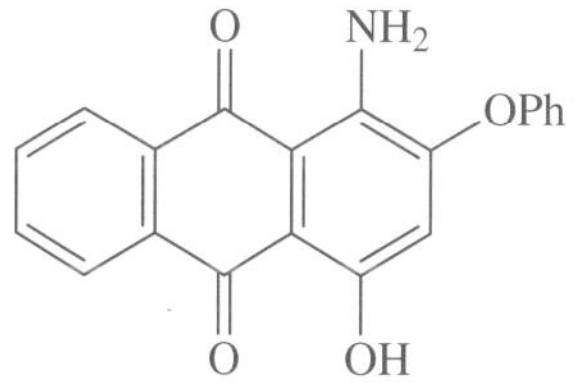

Disperse Red 60

12.4.1 Introduction

Phenanthrene is best represented as a hybrid of the five canonical forms **20–24**. It has a resonance energy of 380 kJ mol^{-1} and is more stable than anthracene. In four of the five resonance structures, the 9,10-bond is double and its length is about the same as an alkenic C=C bond. The numbering system for phenanthrene is shown in **20**. Five different mono-substituted products are possible.

20 **21** **22** **23** **24**

12.4.2 Synthesis of Phenanthrene

There are two major routes to phenanthrene, both of which can be used to prepare substituted derivatives. In the Haworth synthesis (Scheme 12.11), reaction of naphthalene with succinic anhydride yields an oxobutanoic acid which is reduced under Clemmensen conditions to the butanoic acid. Cyclization in sulfuric acid and reduction of the resulting ketone is followed by dehydrogenation over palladium-on-carbon to phenanthrene. Alkyl or aryl derivatives can be obtained by treatment of the intermediate ketone with a Grignard reagent prior to dehydration and oxidation.

In the Pschorr synthesis (Scheme 12.12), a Perkin reaction (see Chapter 6) between 2-nitrobenzaldehyde and sodium phenylacetate in the presence of acetic anhydride yields 3-(2-nitrophenyl)-2-phenylpropenoic acid. Reduction of the nitro group and deamination of the resulting amine *via* its diazonium salt (see Chapter 8) is accompanied by cyclization. Thermal decarboxylation completes the sequence.

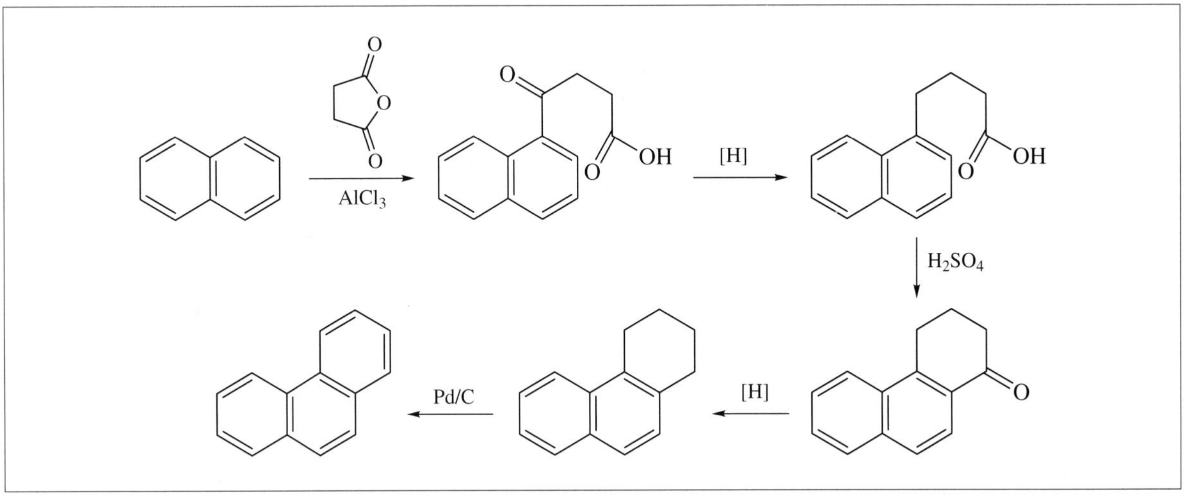

Scheme 12.11

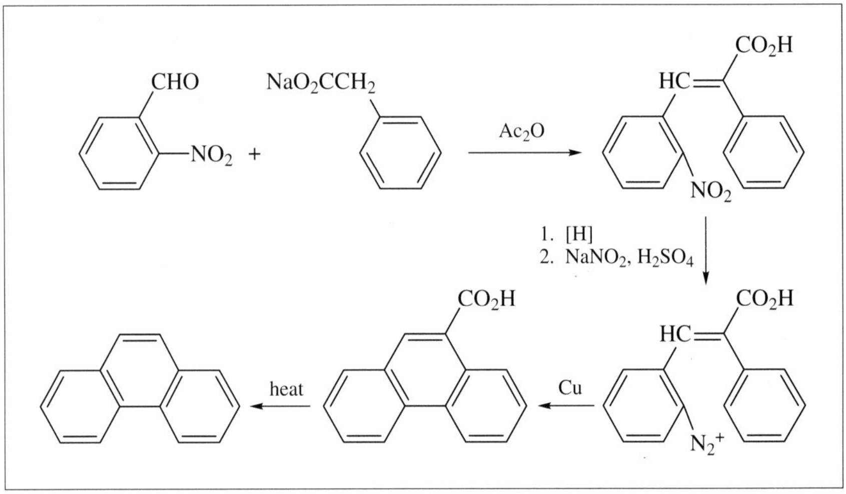

Scheme 12.12

12.4.3 Reactions of Phenanthrene

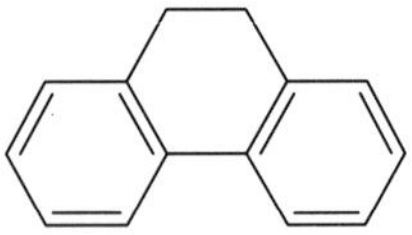

Both reduction (H_2, Pt catalyst) and oxidation (CrO_3, AcOH) of the 9,10-bond are readily accomplished, yielding 9,10-dihydrophenanthrene (**25**) and phenanthra-9,10-quinone (**26**), respectively.

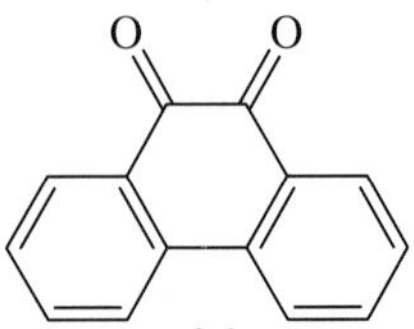

In acetic acid solution, phenanthrene undergoes addition of chlorine to give a mixture of *cis-* and *trans*-9,10-dichloro-9,10-dihydrophenanthrenes (Scheme 12.13). The accompanying formation of an acetoxy derivative suggests that the normal electrophilic addition to a double bond is occurring. The dichloro addition compound eliminates hydrogen chloride to give 9-chlorophenanthrene. However, reaction with bromine in refluxing tetrachloromethane produces 9-bromophenanthrene and an alternative mechanism is probably operating. Other electrophilic substitution reactions of phenanthrene lead to mixtures of products.

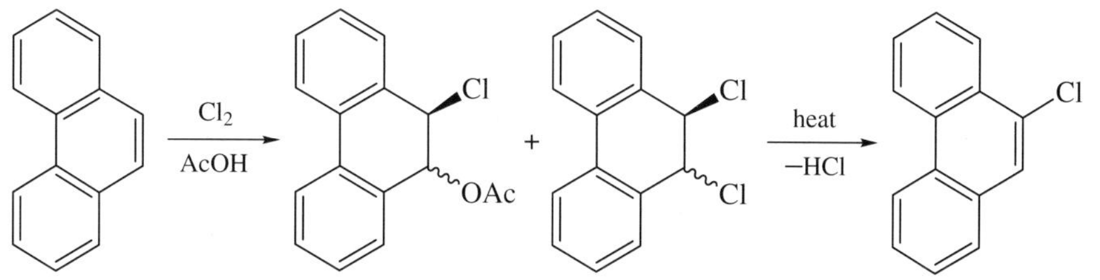

Scheme 12.13

Summary of Key Points

1. Naphthalene, anthracene and phenanthrene can be prepared by an intramolecular Friedel–Crafts reaction in which initially a cyclic anhydride reacts with benzene or naphthalene.

2. Naphthalene and anthracene can also be prepared *via* a Diels–Alder cycloaddition of a quinone with a diene.

3. Electrophilic attack on naphthalene is easier than that on benzene and occurs at the 1-position.

4. Some 1-substituted naphthalene derivatives can be converted into the more thermodynamically stable 2-isomer by heating.

5. Anthracene and phenanthrene are readily oxidized to the quinone and reduced to the dihydro compound.

Problems

12.1. Give the products resulting from the following reactions: (a) 2-naphthol + HCN + $ZnCl_2$ + HCl; (b) anthracene + *cis*-butenedioic anhydride; (c) naphtho-1,4-quinone + hexa-2,4-diene followed by reaction with CrO_3; (d) naphthalene + phthalic anhydride + $AlCl_3$ followed by reaction with polyphosphoric acid and then reduction and dehydrogenation; (e) 1-tetralone + benzaldehyde + NaOH in ethanol.

12.2. Predict the most favourable sites of mononitration for the following compounds and give reasons for your choice: (a) naphthalene-2-carboxylic acid; (b) 2-methylnaphthalene; (c) 1-methylnaphthalene; (d) 1-naphthol; (e) 2-naphthol; (f) 1-nitronaphthalene.

Further Reading

General Organic Chemistry

F. A. Carey, *Organic Chemistry*, 4th edn., McGraw-Hill, New York, 2000.

F. A. Carey and R. J. Sundberg, *Advanced Organic Chemistry*, 4th edn., Plenum, New York, 2001.

J. Clayden, N. Greeves, S. Warren and P. Wothers, *Organic Chemistry*, Oxford University Press, Oxford, 2001.

I. L. Finar, *Organic Chemistry*, vol. 1, 6th edn., Longman, Harlow, 1990.

H. Hopf, *Classics in Hydrocarbon Chemistry*, Wiley, New York, 2000.

R. T. Morrison and R. N. Boyd, *Organic Chemistry*, 6th edn., Prentice-Hall, New York, 1993.

R. O. C. Norman and J. M. Coxon, *Principles of Organic Synthesis*, 3rd edn., Blackie, London, 1993.

M. B. Smith and J. March, *March's Advanced Organic Chemistry*, 5th edn., Wiley, New York, 2000.

T. W. G. Solomons, C. B. Fryhle and R. G. Johnson, *Organic Chemistry*, 7th edn., Wiley, New York, 1999.

Aromaticity and Electrophilic Substitution

P. J. Garratt, *Aromaticity*, Wiley, New York, 1986.

D. Lloyd, *Non-Benzenoid Conjugated Carbocyclic Compounds*, Elsevier, New York, 1984.

D. Lloyd, *The Chemistry of Conjugated Cyclic Compounds*, Wiley, New York, 1989.

R. O. C. Norman and R. Taylor, *Electrophilic Substitution in Benzenoid Compounds*, Elsevier, Amsterdam, 1965.

H. R. Snyder, *Non-Benzenoid Aromatics* (2 vols.), Academic Press, New York, 1969–1971.

R. Taylor, *Electrophilic Aromatic Substitution*, Wiley, New York, 1990.

Organic Reaction Mechanisms

R. Bruckner, *Advanced Organic Chemistry, Reaction Mechanisms*, Academic Press, New York, 2000.

T. H. Lowry and K. S. Richardson, *Mechanism and Theory in Organic Chemistry*, Wesley-Longman, Harlow, 1997.

H. Maskill, *Mechanisms of Organic Reactions*, Oxford Science Publications, Oxford, 1996.

B. Miller, *Advanced Organic Chemistry, Reactions and Mechanisms*, Prentice-Hall, New York, 1997.

P. Sykes, *A Guidebook to Mechanism in Organic Chemistry*, Prentice-Hall, New York, 1996.

Special Topics

C. A. Buehler and D. E. Pearson, *Survey of Organic Syntheses*, Wiley, New York, 1970.

N. Donaldson, *The Chemistry and Technology of Naphthalene Compounds*, Edward Arnold, London, 1958.

B. S. Furness, A. J. Hannaford, P. W. G. Smith and A. R. Tatchell, *Vogel's Textbook of Practical Organic Chemistry*, 5th edn., Longman, Harlow, 1989. This book contains syntheses of a wide selection of aromatic compounds.

A. H. Haines, *Methods for Oxidation of Organic Compounds: Alkanes, Alkenes, Alkynes and Arenes*, Academic Press, London, 1985.

G. A. Olah, *Friedel–Crafts Chemistry*, Wiley, New York, 1973.

P. Powell, *Principles of Organometallic Chemistry*, 2nd edn., Chapman and Hall, London, 1988.

P. N. Rylander, *Hydrogenation Methods*, Academic Press, London, 1985.

K. H. Saunders and R. L. M. Allen, *Aromatic Diazo Compounds*, 3rd edn., Edward Arnold, London, 1985.

R. Stewart, *Oxidation Mechanisms, Application to Organic Chemistry*, Benjamin, New York, 1964.

W. A. Waters, *Mechanisms of Oxidation of Organic Compounds*, Wiley, New York, 1964.

Answers to Problems

Chapter 1

1.1. For the examples cited, the orbitals are arranged as in the following table:

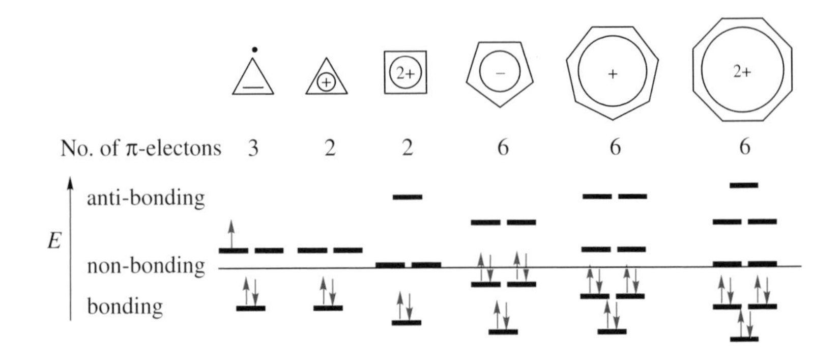

Systems where Huckel's $4n + 2$ rule is obeyed can now easily be seen. All except the cyclopropyl radical can be classed as aromatic species.

1.2. The electronegative oxygen atom polarizes the carbonyl group, attracting electron density from the carbonyl carbon through the π-bond. The resulting canonical form is shown. In this contributing structure, the cyclopropyl cation has two π-electrons and satisfies Hückel's rule, with $n = 0$, and this confers aromatic character on the molecule.

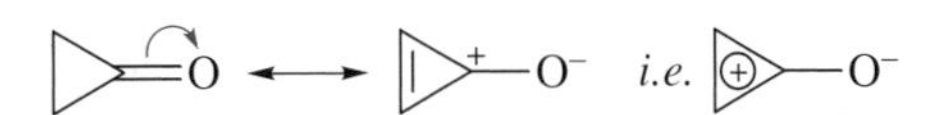

1.3. Thiophene shows significant aromatic properties. The sulfur heteroatom contributes a pair of electrons to the π-cloud and also retains a lone pair of electrons. Furan is also aromatic, though less so than thiophene. The greater electronegativity of oxygen compared with sulfur means there is a greater reluctance to contribute electron density to the π-system. Both compounds resemble pyrrole in the method of achieving aromaticity (see p. 10). The substituted cyclopentadienyl anion is just as aromatic as the unsubstituted anion, but both are best regarded as resonance-stabilized anions. Neither [12]annulene nor cyclononatetraene are aromatic. Both have $4n$ π-electrons. The latter has an sp^3 hybridized carbon atom and so is non-planar.

1.4. (a) 3-Chloro-4-nitroaniline; (b) 4-bromo-3-methylbenzenesulfonic acid; (c) 3-amino-5-nitrobenzoic acid; (d) 2,4-dichloro-3-ethylphenol; (e) 2-chloro-5-methylbenzoic acid.

1.5.

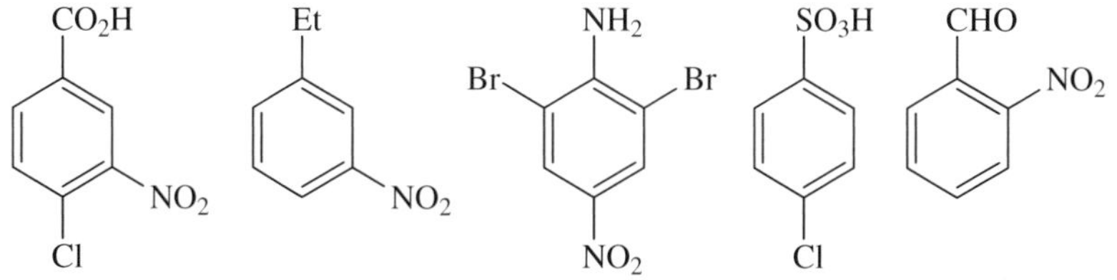

CO_2H Et NH_2 SO_3H CHO

Br Br NO_2

NO_2 NO_2

Cl NO_2 Cl

(a) (b) (c) (d) (e)

Chapter 2

2.1. (a) Benzene reacts with benzoyl chloride under Friedel–Crafts conditions to produce benzophenone.
(b) Acetanilide is nitrated with mixed acid in the *ortho* and *para* positions and therefore a mixture of the two products results. The major product is the 4-isomer because the bulk of the NHCOMe group hinders attack at the 2-position.
(c) In bromobenzene, the bromine atom deactivates the ring to electrophilic substitution, but directs substitution to the *ortho* and *para* positions. The products are 2-bromobenzenesulfonic acid and the 4-isomer.
(d) In 1-methoxy-2-nitrobenzene, the directing effects of the two substituents are working together. The methoxy group is activating and directs *ortho*/*para*; the nitro group is deactivating at all positions, but directs *meta*. There is steric hindrance to attack *ortho* to the methoxy group and so nitration with mixed acid gives predominantly 1-methoxy-2,4-dinitrobenzene.

(e) The trifluoromethyl group is deactivating and *meta* directing and hence the product is 1-bromo-3-(trifluoromethyl)benzene.

2.2. (a) 3-Bromoaniline may conveniently be synthesized by nitration of benzene followed by bromination in the presence of iron powder and reduction of the nitro group.
(b) The methyl group is *ortho*/*para* directing and so the synthesis of 3-nitrobenzoic acid from toluene starts by oxidation of the methyl group to give the *meta*-directing carboxylic acid group. Nitration of the resulting benzoic acid gives the required product.
(c) Nitration of toluene produces a mixture of 2- and 4-nitrotoluenes. Separation of the 4-isomer followed by oxidation of the methyl group produces 4-nitrobenzoic acid. Compare the answers to parts (b) and (c) to see how manipulation of functional groups enables different isomers to be obtained.
(d) Treatment of benzoic acid with thionyl chloride produces benzoyl chloride, reaction of which with toluene in the presence of aluminium chloride gives 4-methylbenzophenone. Steric considerations will limit substitution in the 2-position of toluene.
(e) Dinitration of chlorobenzene gives 1-chloro-2,4-dinitrobenzene. The presence of the two strongly electron-withdrawing groups activates the chlorine atom to nucleophilic substitution and thus reaction with sodium methoxide produces 1-methoxy-2,4-dinitrobenzene.

2.3. (a) The initial Friedel–Crafts alkylation with an alkyl halide introduces an electron-donating alkyl group and thus the product is more reactive than the original reactant and undergoes further alkylation so that mixtures result. This is not the case with Friedel–Crafts acylation because the acyl group is electron-withdrawing and deactivating.
(b) Monobromination of ethyl 4-methylbenzoate gives one product, ethyl 3-bromo-4-methylbenzoate, because the electronic influences of the two substituents combine to direct attack to the 3-position. This position is *ortho* to the activating methyl group and *meta* to the deactivating ester group. The other site activated by the methyl group, the *para* position, is blocked by the ester function.
(c) The two products of this reaction, 3-aminotoluene and 4-aminotoluene, are the result of the intermediacy of an aryne. Abstraction of a proton by a strong base is accompanied by loss of Cl^- (overall loss of HCl), which produces a "triple" bond between the 3- and 4-positions. The subsequent addition of ammonia occurs equally at

the 3- and 4-positions and produces a 1:1 mixture of the two products.
(d) The primary carbocationic complex $[Me_2CHCH_2^+][AlCl_4^-]$ produced from 1-chloro-2-methylpropane on reaction with $AlCl_3$ rearranges to the more stable tertiary cation $[Me_3C^+][AlCl_4^-]$ prior to reaction with benzene. The product is therefore *tert*-butylbenzene and not the expected isobutylbenzene.

Chapter 3

3.1. (a) Propan-2-ol is dehydrated by sulfuric acid to propene, which is protonated by the acid to generate the carbocation Me_2CH^+. A Friedel–Crafts alkylation follows and, since the chlorine atom of chlorobenzene is *ortho*/*para* directing, the major product is 1-chloro-4-isopropylbenzene.
(b) This is an example of a Friedel–Crafts acylation. The bromine substituent directs attack to the *ortho* and *para* positions. Steric factors hinder attack at the 2-position and the major product is 4-bromoacetophenone.
(c) Another Friedel–Crafts acylation. The product is benzyl phenyl ketone, $PhCH_2COPh$.
(d) Boron trifluoride etherate, $BF_3{\cdot}Et_2O$, is the catalyst in this Friedel–Crafts acylation by acetic anhydride. Attack is possible in both the 2- and the 4-positions, but steric considerations suggest that the 4-isomer will predominate.
(e) The acylating reagent is a cyclic anhydride and, unlike in part (d) when acetic acid was a by-product, the leaving group here remains attached to the product, which is 4-oxo-4-phenylbutanoic acid.

3.2. (a) Toluene reacts with chlorine in the presence of iron(III) chloride to give a mixture of 2- and 4-chlorotoluenes from which the latter is separated. The methyl group is oxidized to produce 4-chlorobenzoic acid, which is then nitrated using mixed acid to give 4-chloro-3-nitrobenzoic acid. Both substituents direct attack to the 3-position.
(b) Toluene is mononitrated and the 4-nitrotoluene separated from admixture with the 2-isomer. Chlorination ($Cl_2/FeCl_3$) is directed to the 2-position by both substituents and oxidation of the methyl group then produces 2-chloro-4-nitrobenzoic acid. Parts (a) and (b) provide a further example of manipulation of functional groups to allow the preparation of different isomers
(c) (Chloromethyl)benzene is reacted with toluene in the presence

of $AlCl_3$ to give 4-benzyltoluene and the methyl group is then oxidized to the carboxylic acid.
(d) Benzene is acylated under Friedel–Crafts conditions to produce acetophenone. Nitration gives 3-nitroacetophenone and Wolff–Kishner reduction yields 1-ethyl-3-nitrobenzene.
(e) Benzene is acylated with 2-methylpropanoyl chloride and the carbonyl group in the product, 2-methyl-1-phenylpropanone, is reduced by Zn(Hg) and HCl (Clemmensen reduction) to produce isobutylbenzene. Direct alkylation with 1-chloro-2-methylpropane results in rearrangement of the alkyl carbocation and formation of *tert*-butylbenzene and so indirect alkylation has to be used [see Problem 2.3(d)].

Chapter 4

4.1. The acidity of substituted phenols is dependent on the electronic properties of the substituent. An electron-donating group decreases the acidity and an electron-withdrawing group increases the acidity. The position of the substituent relative to the hydroxyl group is important since this controls whether a mesomeric or an inductive effect or both are operative.
(a) 4-Cyanophenol (–M/–I) is more acidic than 3-cyanophenol (only –I).
(b) 4-Chlorophenol (weak +M/strong –I) is more acidic than 4-methoxyphenol (strong +M/weak –I).
(c) 2,4-Dinitrophenol (–M/–I) is a stronger acid than 3,5-dinitrophenol (only –I).
(d) 4-Nitrophenol (strong –M) is a stronger acid than 4-hydroxybenzaldehyde (CHO has a weaker –M effect).

4.2. (a) $OSO_2C_6H_4Me$–4 (b) OCH_2CO_2H Cl Cl Cl (c) OH $CH{=}CH_2$ CH Et

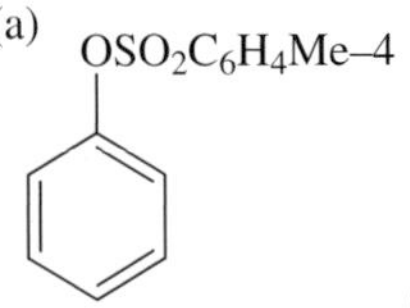

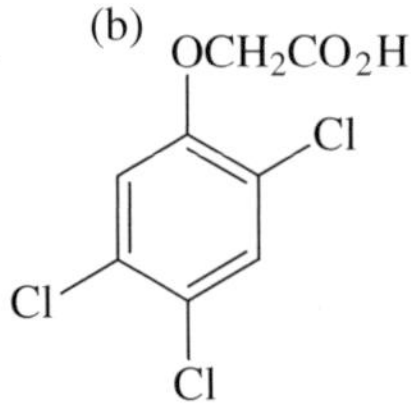

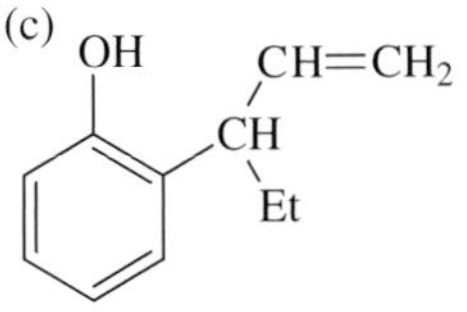

(d) OH COMe Me (e) OMe $COCH_2CH_2CO_2H$

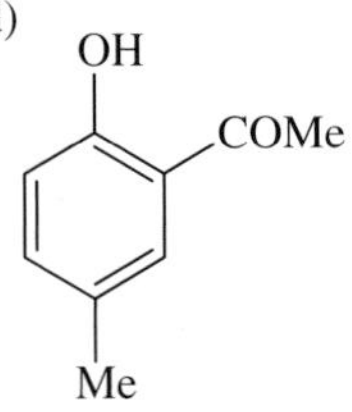

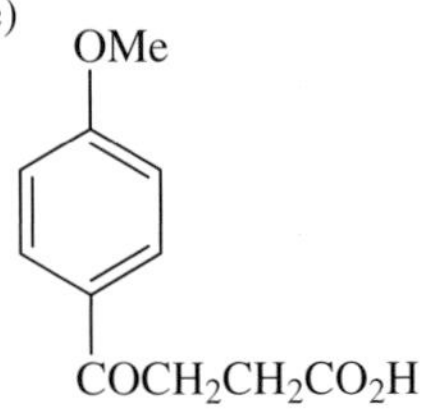

4.3. (a) Nitrate benzenesulfonic acid in the 3-position, fuse the product with sodium hydroxide to displace the sulfonic acid with hydroxyl, and alkylate the hydroxyl group with methyl iodide to form the methoxy group.
(b) Reaction of phenol under Kolbe–Schmidt conditions with carbon dioxide under pressure gives 2-hydroxybenzoic acid, which can be esterified in methanol using an acid catalyst such as sulfuric acid.
(c) Nitrate toluene to produce a mixture of 2- and 4-nitrotoluenes and separate the 4-isomer, which is reduced to 4-aminotoluene. After monobromination *ortho* to the amino group, the new aniline derivative is converted to the phenol by diazotization and reaction with boiling water.

4.4. The three strongly electron-withdrawing nitro groups (–M/–I) make 2,4,6-trinitrophenol much more acidic than phenol. It is a sufficiently strong acid to react with the weakly basic sodium hydrogen carbonate, behaving as a carboxylic acid. This reaction is used to distinguish carboxylic acids from phenols. The stronger base sodium hydroxide will react with both of these phenols.

Chapter 5

5.1. (a) Nucleophilic attack on the carbonyl function results in cyclization. Proton transfer followed by loss of water gives the product:

O
ÖH OH
Ö⁻
O
$\overset{+}{O}H_2$
O O

(b) Nucleophilic attack on the electron-deficient carbon atom of the nitrile group leads to a hydroxyimine, which tautomerizes to an amide. Further nucleophilic attack at the carbonyl group results in hydrolysis of the amide to the salt of the carboxylic acid and ammonia:

HŌ: C≡N:
HO—C=N̄:
H_2O
HO—C=NH
O=C—NH_2

$$\text{Ph—C(=O)NH}_2 + \text{HO}^- \longrightarrow \left[\text{Ph—C(O}^-\text{)(OH)NH}_2 \xrightarrow{-\text{NH}_2^-} \text{Ph—C(=O)OH}\right] \longrightarrow \text{Ph—C(=O)O}^- + \text{NH}_3$$

5.2. The acidity of substituted benzoic acids is dependent on the relative ability of the substituent to donate or withdraw electron density. The more strongly a group withdraws electrons, the greater is the acidity. Thus the order of increasing acid strength is:
4-hydroxybenzoic acid < benzene-1,4-dicarboxylic acid < 4-acetylbenzoic acid < 4-cyanobenzoic acid

5.3.
(a) Nitrate; separate 4-nitrotoluene; chlorinate methyl group (Cl_2, *hν*) to $-CH_2Cl$; treat with KCN to give $-CH_2CN$; hydrolyse to $-CH_2CO_2H$ to give (4-nitrophenyl)acetic acid.
(b) Nitrate; reduce 4-nitrotoluene to 4-methylaniline; acetylate amine to –NHAc; brominate (Br_2, AcOH) in the 2-positon; hydrolyse to 2-bromo-4-methylaniline; deaminate by diazotization ($NaNO_2$, HCl) and treatment with H_3PO_2, giving 3-bromotoluene; form Grignard and treat with CO_2 to give 3-methylbenzoic acid.
(c) Sulfonate to give toluene-4-sulfonic acid; oxidize methyl group ($KMnO_4$) to give 4-sulfobenzoic acid.
(d) Dibrominate to 2,4-dibromotoluene; oxidize methyl group to $-CO_2H$; form acid chloride ($SOCl_2$); treat with ammonia to give 2,4-dibromobenzamide.

Chapter 6

6.1. (a) A carbanion is generated from the methyl group of nitromethane, activated by the nitro group. Condensation with benzaldehyde produces the nitrostyrene.

$$O_2NCH_2\text{—H} + OH^- \rightleftharpoons O_2NCH_2^- + H_2O$$

$$\text{PhCH=O} + O_2NCH_2^- \rightleftharpoons \text{PhCH(O}^-\text{)CH}_2NO_2 \rightleftharpoons \text{PhCH(OH)CH}_2NO_2 \longrightarrow \text{PhCH=CHNO}_2$$

(b) The carbanion generated from the ester reacts preferentially at the aldehyde carbon atom since this is more electrophilic than the ester carbonyl unit:

$$PhCH_2CO_2Et + OEt^- \rightleftharpoons Ph\bar{C}HCO_2Et + EtOH$$

$$PhCH{=}O + Ph\bar{C}HCO_2Et \rightleftharpoons \underset{PhCHCO_2Et}{PhCH{-}O^-} \rightleftharpoons \underset{PhCHCO_2Et}{PhCH{-}OH} \longrightarrow \underset{PhCCO_2Et}{PhCH} \ (\text{C=C})$$

(c) This is an example of a crossed (or mixed) aldol condensation. The carbanion derived from acetophenone reacts with benzaldehyde rather than undergoing a self-condensation. The latter reaction is kept to a minimum by ensuring that there is only a low concentration of carbanion by adding the acetophenone to a mixture of benzaldehyde and the base. The former reaction is favoured because an aldehyde is more electrophilic than a ketone.

$$PhCOCH_2{-}H \ {:}OH^- \underset{}{\overset{-H_2O}{\rightleftharpoons}} PhC({=}O){-}\ddot{C}H_2^- \longleftrightarrow PhC(O^-){=}CH_2$$

$$PhC(H){=}O \ \ \bar{\ddot{C}}H_2COPh \rightleftharpoons PhC(O^-)(H){-}CH_2COPh \overset{H_2O}{\rightleftharpoons} PhCH(OH){-}CH_2COPh \xrightarrow{-H_2O} PhCH{=}CHCOPh$$

(d) The nucleophilic amino group attacks the electron-deficient carbonyl carbon atom of benzaldehyde. Elimination of water gives an imine (a Schiff base):

$$PhCH{=}O + Ph\ddot{N}H_2 \longrightarrow \underset{PhNH_2^+}{PhCH{-}O^-} \longrightarrow \underset{PhNH}{PhCHOH} \xrightarrow{-H_2O} PhCH{=}NPh$$

6.2. (a) This is an example of the Cannizzaro reaction and produces 4-chlorobenzoic acid and 4-chlorobenzyl alcohol.
(b) Formaldehyde has no α-hydrogen atom and so cannot form a carbanion. However, it is very reactive towards carbanions and reacts readily with that derived from acetophenone to give the alcohol shown, which may dehydrate:

$$H_2C{=}O + PhCO\ddot{C}H_2^- \rightleftharpoons PhCOCH_2CH_2OH \ [\longrightarrow PhCOCH{=}CH_2]$$

(c) Nucleophilic addition of NH_2OH to the carbonyl group followed by elimination of water leads to the oxime **1**.

(d) A carbanion can be generated from cyclohexanone on treatment with a base. The carbanion reacts preferentially with benzaldehyde to give **2**.

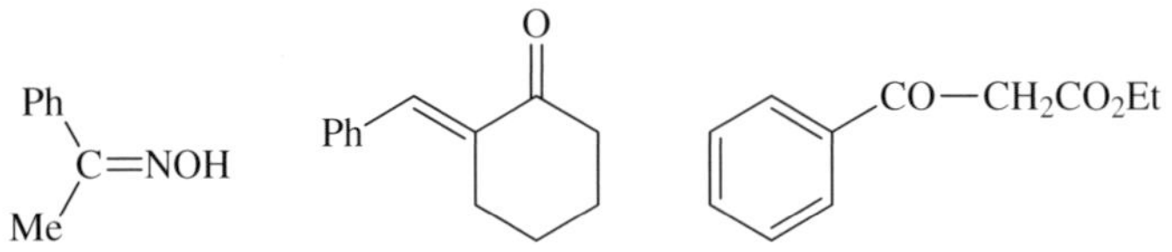

Ph(Me)C=NOH **1**

O, Ph **2**

$CO-CH_2CO_2Et$ **3**

(e) This is an example of a crossed Claisen condensation. Unlike ethyl acetate, ethyl benzoate possesses no α-hydrogen atom and so cannot form a carbanion. Here, $:\bar{C}H_2CO_2Et$, derived from $MeCO_2Et$, reacts with the ester function of ethyl benzoate to give ethyl benzoylacetate (**3**).

It is possible that some ethyl acetoacetate (ethyl 3-oxobutanoate), $MeCOCH_2CO_2Et$, will also be formed through the self-condensation of ethyl acetate, but this reaction can be suppressed as described in Problem 6.1(c) above.

6.3. (a) Reduction of 4,4′-dimethoxybenzophenone with $NaBH_4$.
(b) A crossed aldol condensation between phenylacetaldehyde, $PhCH_2CHO$, the source of the carbanion, and benzaldehyde.
(c) A Perkin reaction between 4-nitrobenzaldehyde and acetic anhydride (the carbanion source) using sodium acetate as the catalyst.
(d) A crossed aldol condensation between benzaldehyde and acetone.

Chapter 7

7.1. The ease of electrophilic substitution is influenced by the electronic behaviour of the substituents. Donor groups help the reaction and direct attack to the *ortho* and *para* positions; electron-withdrawing groups make attack, which occurs at the *meta* position, more difficult. Thus the reactivity of the compounds is:
(a) Methoxybenzene > benzene > (trifluoromethyl)benzene > benzonitrile. The major products are 1-methoxy-4-nitrobenzene, nitrobenzene, 1-nitro-3-(trifluoromethyl)benzene and 3-nitrobenzonitrile.
(b) Acetanilide > chlorobenzene > benzoic acid > acetophenone. The major products are 4-nitroacetanilide, 1-chloro-4-nitrobenzene, 3-nitrobenzoic acid and 3-nitroacetophenone.
(c) Phenol > bromobenzene > ethyl benzoate > nitrobenzene. The

major products are 4-nitrophenol, 1-bromo-4-nitrobenzene, ethyl 3-nitrobenzoate and 1,3-dinitrobenzene.

7.2. (a) This compound is prepared by the Friedel–Crafts reaction of propanoyl chloride with benzene in the presence of aluminium chloride, followed by nitration in the 3-position.
(b) Nitration of benzene produces nitrobenzene, but direct introduction of a second nitro group would occur at the *meta* position. After reduction of the nitro group and acetylation of the amine, nitration now takes place at the *para* position, leading to 4-nitroacetanilide. Hydrolysis, diazotization and conversion of the diazonium group into a nitro group (see Chapter 8) gives the product. Direct oxidation of the amino group using H_2SO_5 or CF_3CO_2H is an alternative step.
(c) Aniline → acetanilide; sulfonate in the 4-position; nitrate to give 4-(acetylamino)-3-nitrobenzenesulfonic acid; desulfonate and hydrolyse to form the target molecule.

Chapter 8

8.1. (a) The basicity increases in the order: diphenylamine < aniline < benzylamine. The relative ability of the lone pair of electrons on the nitrogen atom to interact with the π-electron cloud of the aromatic ring affects the basicity. The aliphatic amino group in benzylamine cannot interact with the π-cloud of the benzene ring and so this is the strongest base. Diphenylamine, in which the amino function is attached to two aromatic rings, is the weakest.
(b) Electron-withdrawing groups decrease the basicity of aniline and therefore the stronger the electron-withdrawing ability of the group, the weaker the base. Thus the order of increasing basicity is: 4-aminobenzaldehyde < methyl 4-aminobenzoate < 4-bromoaniline.

8.2. (a) Aniline is treated with chlorine water to produce 2,4,6-trichloroaniline and this product is diazotized and deaminated by treatment with phosphinic acid.
(b) Aniline is acetylated and reacted with MeI under Friedel–Crafts conditions to introduce the methyl group and the acetyl group is removed to give 4-methylaniline. Diazotization and reaction with copper(I) chloride produces 4-chlorotoluene and oxidation of the methyl group gives 4-chlorobenzoic acid.
(c) Aniline is acetylated and nitrated. The 4-nitroacetanilide is isolated and hydrolysed to regenerate the amino group, which is then

diazotized and reacted with copper(I) chloride to give 1-chloro-4-nitrobenzene.
(d) Benzene is nitrated and brominated to give 1-bromo-3-nitrobenzene. Reduction to the amine is followed by diazotization and formation of the diazonium tetrafluoroborate. On heating, 1-bromo-3-fluorobenzene is produced.
(e) Nitrate benzene and chlorinate it to give 1-chloro-3-nitrobenzene. Reduction and acetylation of the amino group gives 3-chloroacetanilide.

8.3. (a) 4-Bromoanilinium chloride (the hydrochloride salt); (b) 1,4-dibromobenzene; (c) 4-bromoacetanilide; (d) 4-bromo-*N*,*N*,*N*-triethylanilinium iodide; (e) 2,4,6-tribromoaniline.

8.4. On heating benzenediazonium-2-carboxylate, CO_2 and N_2 are eliminated and benzyne is formed. Furan traps the benzyne as *cis*-1,4-dihydronaphthalene-1,4-endoxide, which is hydrolysed by HCl to 1-naphthol.

Chapter 9

9.1. (a) 1-Methoxy-4-nitrobenzene; (b) 2-chloro-1-methoxy-4-nitrobenzene; (c) indane-1-nitrile (the base generates both an aryne and a carbanion; intramolecular attack by the latter on the former results in cyclization to the indane); (d) 3-amino-4-methylbiphenyl and 2-amino-4-methylbiphenyl.

9.2. (a) Chlorinate benzene (Cl_2, $FeCl_3$) and then nitrate the chlorobenzene and separate the 4-isomer.
(b) Nitrate benzene and brominate (Br_2, $FeBr_3$).
(c) Sulfonation of benzene followed by fusion with sodium hydroxide produces phenol. This is then methylated and the methoxybenzene nitrated and the 4-isomer reduced, diazotized and treated with copper(I) cyanide to give 4-methoxybenzonitrile.
(d) Fluorobenzene may be prepared by nitration of benzene followed by reduction, diazotization and introduction of fluorine *via* reaction with tetrafluoroborate ion.

9.3. (a) Nucleophilic substitution of the activated chlorine atom by hydrazine leads to 2,4-dinitrophenylhydrazine, compound X. This product is a standard reagent for the detection of carbonyl compounds, with which it forms a yellow/orange precipitate (Z) of the 2,4-dinitrophenylhydrazone:

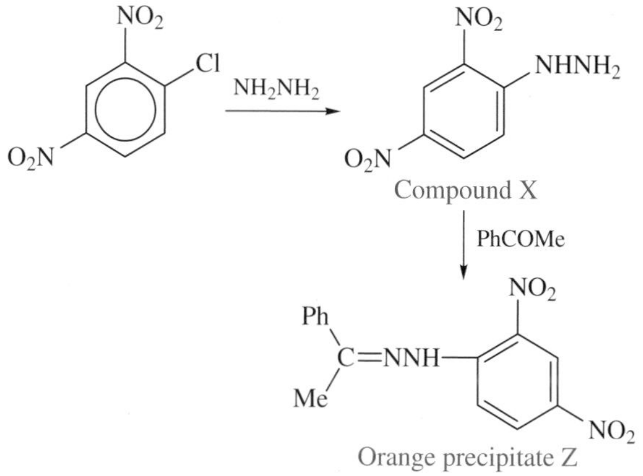
NO2
Cl
O2N
NH2NH2
NO2
NHNH2
O2N
Compound X
PhCOMe
NO2
Ph
C=NNH
Me
NO2
Orange precipitate Z

(b) Friedel–Crafts acylation of bromobenzene gives a mixture of 2- and 4-bromoacetophenones in which the latter predominates (A). The ketone function forms an oxime (B) on treatment with hydroxylamine. Under acidic conditions, oximes undergo a stereospecific Beckmann rearrangement in which the group *anti* to the hydroxy group migrates. The product results from migration of the aryl group and is 4-bromoacetanilide (D):

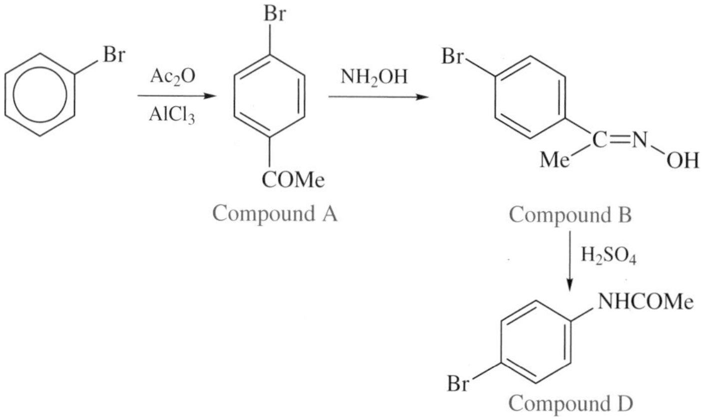
Br
Ac2O
AlCl3
Br
COMe
Compound A
NH2OH
Br
C=N
Me
OH
Compound B
H2SO4
NHCOMe
Br
Compound D

(c) The fluorine atom in 1-fluoro-2,4-dinitrobenzene is activated towards nucleophilic attack by the two nitro group. The free amino group of the N-terminal amino acid of the peptide is nucleophilic and displaces the fluorine atom. Subsequent hydrolysis of the product gives the free amino acids that made up the peptide, except for the terminal amino acid which remains linked to the dinitrophenyl unit. This product G is yellow in colour. The terminal amino acid has been labelled by the aryl group:

NO_2 F O_2N

+ $H_2NCHCONHCHCO$—polypeptide (Ph, Me)

↓

NO_2 O_2N NHCHCONHCHCO—polypeptide (Ph, Me)

↓ hydrolyse

NO_2 O_2N NHCHCO$_2$H (Ph) + H_2NCHCO_2H (Me) + other amino acids

Yellow compound G

Chapter 10

10.1. Using palladium catalysis:

(a) Br → (MeHN–CH$_2$Ph; Pd^0, BINAP, KOBu^t) → Me–N

Using arene–chromium tricarbonyl chemistry:

(b) Br → (naphthalene–Cr(CO)$_3$) → Br, Cr(CO)$_3$ → (PrOH) → O

10.2. (a) The synthesis uses a Suzuki coupling and involves the sequence (i) oxidative addition, (ii) transmetallation and (iii) reductive elimination with regeneration of the palladium catalyst:

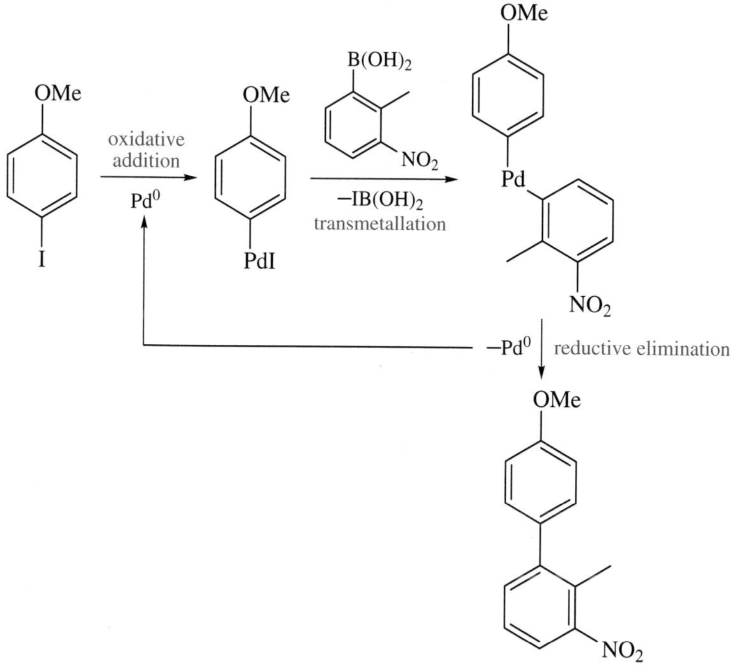

(b) Directed *o*-lithiation and electrophilic attack by iodine are followed by a Heck reaction. The mechanism involves oxidative addition, π-complexation and β-hydride elimination:

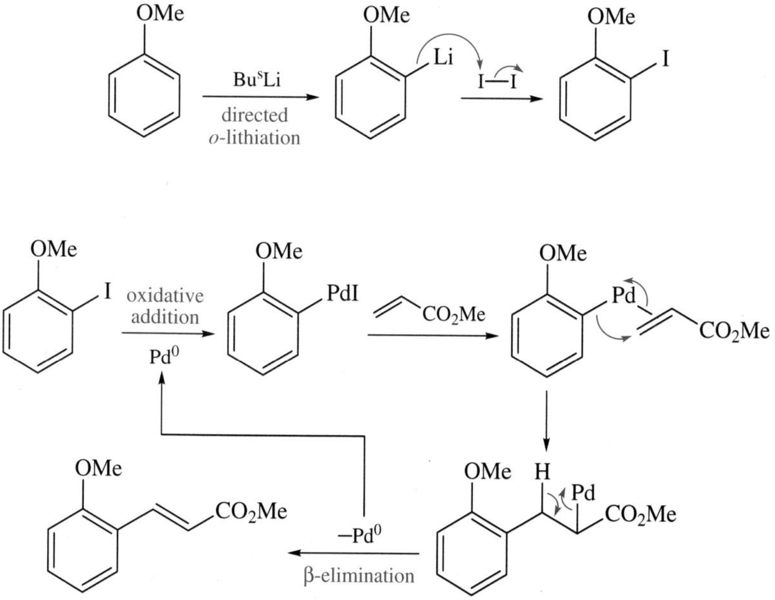

10.3. (a) Conversion of iodobenzene into the Grignard or phenyllithium compound, then reaction with butan-2-one followed by hydrolysis gives the product.
(b) Reaction of phenyllithium with 4-methylbenzonitrile followed by hydrolysis gives the product.
(c) Reaction of phenyllithium with carbon dioxide followed by hydrolysis and nitration of the benzoic acid produced yields the target compound.
(d) Reaction of phenyllithium with butanal and hydrolysis gives 1-phenylbutan-1-ol.

10.4. (a) Nitrobenzene is reduced to aniline and converted to iodobenzene *via* diazotization and reaction with aqueous potassium iodide solution. Conversion into phenylmagnesium iodide and reaction with solid carbon dioxide in diethyl ether followed by hydrolysis gives benzoic acid.
(b) Phenol is methylated and then monobrominated to give 4-bromoanisole (1-bromo-4-methoxybenzene). This is reacted with magnesium turnings and then with carbon dioxide. Nitration of the resulting 4-methoxybenzoic acid produces the required product.
(c) Monobromination of toluene followed by conversion to the Grignard reagent, then reaction with benzophenone or two moles of methyl benzoate followed by hydrolysis gives the product alcohol.
(d) Reaction of benzoic acid with thionyl chloride gives the acid chloride, which with ammonia yields the amide. Dehydration produces benzonitrile which reacts with propylmagnesium bromide to produce 1-phenylbutan-1-one after hydrolysis.

Chapter 11

11.1.

OMe (anisole) $\xrightarrow[\text{Bu}^t\text{OH}]{\text{Na/NH}_3}$ OMe (1-methoxycyclohexa-1,4-diene)

$CONMe_2$ (benzene) $\xrightarrow[\text{2. PhCH}_2\text{Br}]{\text{1. Na/NH}_3}$ Me_2NOC, CH_2–Ph (cyclohexa-2,5-diene)

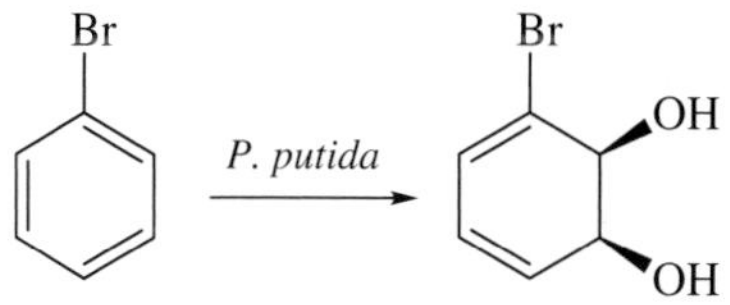

Br P. putida Br OH OH

Formed as a single enantiomer

11.2.

(a) A 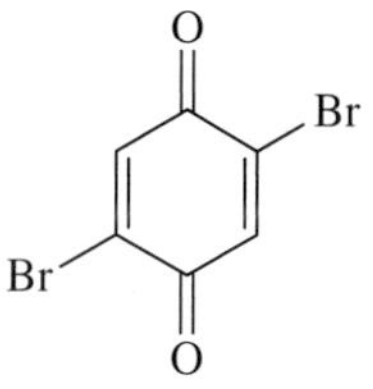 B 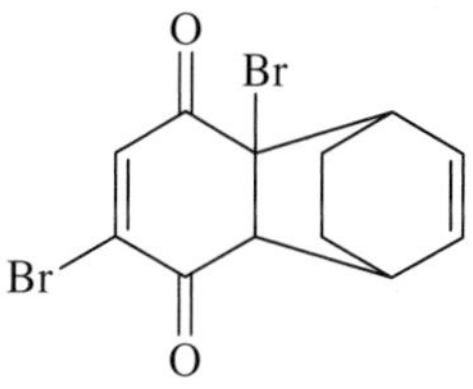 C

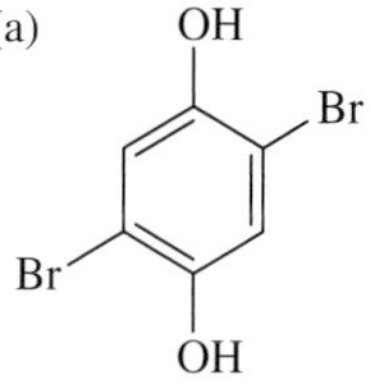

(b) F 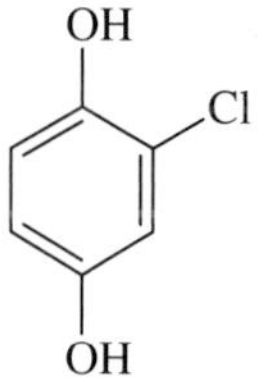 G (c) M

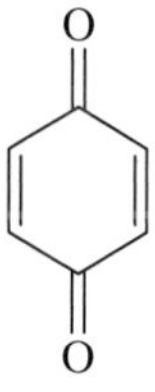

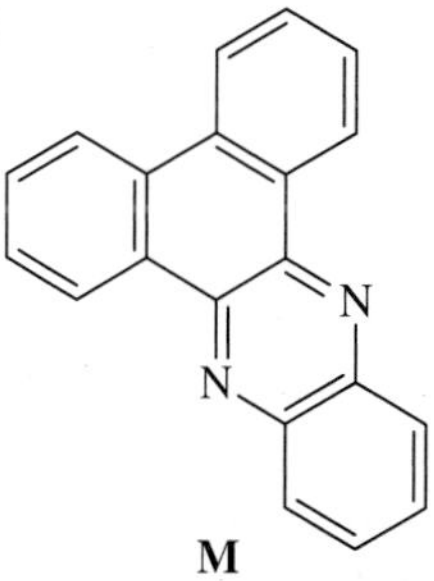

Chapter 12

12.1. (a) This is an example of the Gattermann formylation. The attacking electrophile is $HC{\equiv}NH^+$ and the initial product is an iminium chloride, hydrolysis of which gives the aldehyde:

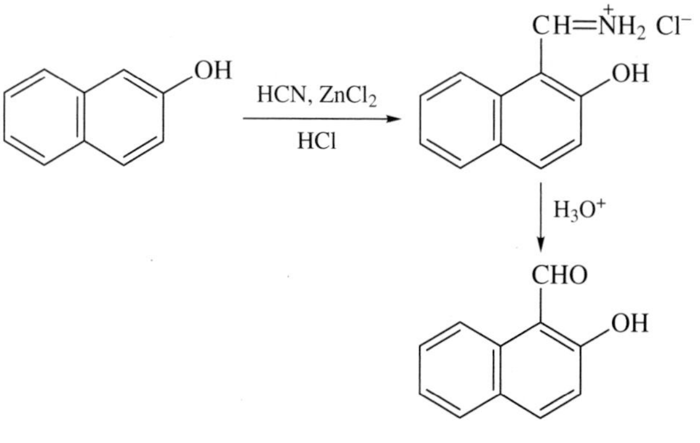

HCN, $ZnCl_2$ HCl → $CH{=}\overset{+}{N}H_2\ Cl^-$ OH → H_3O^+ → CHO OH

(b) Anthracene reacts readily with dienophiles across the 9,10-positions because very little resonance stabilization energy is lost in the process. *cis*-Butenedioic anhydride gives the adduct:

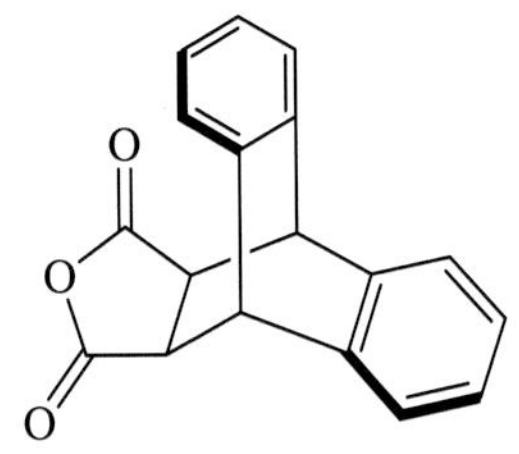

O
O
O

(c) Naphtho-1,4-quinone behaves as a dienophile with dienes and the initial product is a reduced anthraquinone. Oxidation gives the anthra-9,10-quinone:

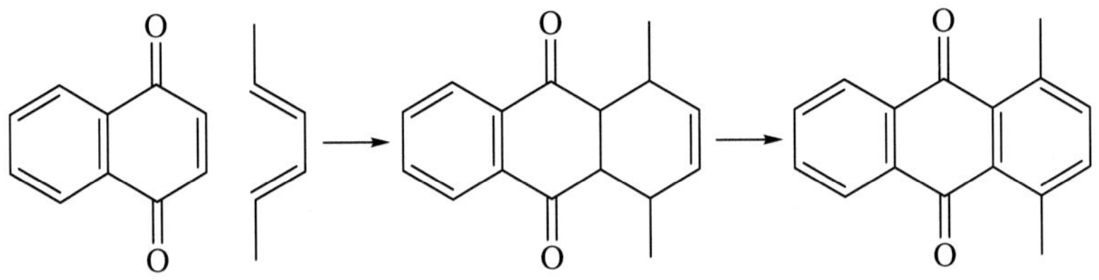

O O O
O O O

(d) An initial Friedel–Crafts acylation is followed by cyclization (an intramolecular Friedel–Crafts acylation) to an anthraquinone derivative. Various reducing agents convert anthraquinones directly to anthracene and the final product here is the tetracyclic compound benz[*a*]anthracene:

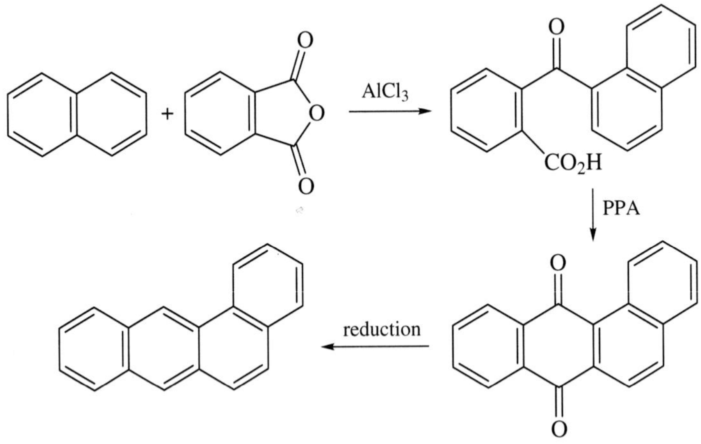

+ O O O $AlCl_3$ O CO_2H

PPA

reduction O O

(e) The carbanion derived from 1-tetralone reacts with the carbonyl group of benzaldehyde in a crossed aldol condensation. The final dehydration ensures that the reaction proceeds to completion:

12.2. Generally, when an electron-withdrawing group is present in naphthalene, electrophilic attack occurs at the α-positions of the other ring. Naphthalenes containing an electron-donating substituent are attacked at the *ortho* and *para* positions of the same ring. (a) 5-Nitronaphthalene-2-carboxylic acid and the 8-nitro isomer; (b) 2-methyl-1-nitronaphthalene; (c) 1-methyl-4-nitronaphthalene; (d) 4-nitro-1-naphthol; (e) 1-nitro-2-naphthol; (f) 1,5- and 1,8-dinitronaphthalenes.

Subject Index